Springer Series in
MATERIALS SCIENCE 77

Springer Series in
MATERIALS SCIENCE

Editors: R. Hull   R. M. Osgood, Jr.   J. Parisi   H. Warlimont

The Springer Series in Materials Science covers the complete spectrum of materials physics, including fundamental principles, physical properties, materials theory and design. Recognizing the increasing importance of materials science in future device technologies, the book titles in this series reflect the state-of-the-art in understanding and controlling the structure and properties of all important classes of materials.

P.O. Holtz   Q.X. Zhao

Impurities Confined in Quantum Structures

With 60 Figures

 Springer

Prof. Dr. Per Olof Holtz
Linköping University, Department of Physics and Measurement Technology, Materials Science
58183 Linköping, Sweden
E-mail: poh@ifm.liu.se

Dr. Qing Xiang Zhao
Chalmers University of Technology, Department of Physics, Physical Electronics and Photonics
41296 Göteborg, Sweden
E-mail: Zhao@fy.chalmers.se

Series Editors:

Professor Robert Hull
University of Virginia
Dept. of Materials Science and Engineering
Thornton Hall
Charlottesville, VA 22903-2442, USA

Professor R. M. Osgood, Jr.
Microelectronics Science Laboratory
Department of Electrical Engineering
Columbia University
Seeley W. Mudd Building
New York, NY 10027, USA

Professor Jürgen Parisi
Universität Oldenburg, Fachbereich Physik
Abt. Energie- und Halbleiterforschung
Carl-von-Ossietzky-Strasse 9–11
26129 Oldenburg, Germany

Professor Hans Warlimont
Institut für Festkörper-
und Werkstofforschung,
Helmholtzstrasse 20
01069 Dresden, Germany

ISSN 0933-033X

ISBN 3-540-22320-7 Springer Berlin Heidelberg New York

Library of Congress Control Number: 2004107649

Springer is a part of Springer Science+Business Media.

springeronline.com

© Springer-Verlag Berlin Heidelberg 2004
Printed in Germany

Typesetting by the editors
Production: PTP-Berlin Protago-TeX-Production GmbH, Berlin
Cover concept: eStudio Calamar Steinen
Cover production: *design & production* GmbH, Heidelberg

Printed on acid-free paper   SPIN: 10636049   57/3141/YU   5 4 3 2 1 0

Preface

The dramatic impact of low dimensional semiconductor structures on current and future device applications cannot be overstated. Research over the last decade has highlighted the use of quantum engineering to achieve previously unknown limits for device performance in research laboratories. The modified electronic structure of semiconductor quantum structures results in transport and optical properties, which differ from those of constituent bulk materials. The possibility to tailor properties, such as bandgap, strain, band offset etc., of two-dimensional (2D) semiconductors, e.g. quantum wells, for specific purposes has had an extensive impact on the electronics, which has resulted in a dramatic renewal process. For instance, 2D structures are today used in a large number of high speed electronics and optoelectronic applications (e.g. detectors, light emitting diodes, modulators, switches and lasers) and in daily life, in e.g. LED-based traffic lights, CD-players, cash registers.

The introduction of impurities, also in very small concentrations, in a semiconductor can change its optical and electrical properties entirely. This attribute of the semiconductor is utilized in the manifoldness of their applications. This fact constitutes the principal driving force for investigation of the properties of the impurities in semiconductors. While the impurities in bulk materials have been investigated for a long time, and their properties are fairly well established by now, the corresponding studies of impurities in quantum wells is a more recent research area. The reduction in dimensionality and symmetry for a confined defect and the effect of the confinement on the electronic particles constitute the background for the great deal of attention for this kind of structure.

In this book, the major progress on the investigations of impurities confined in quantum wells is reviewed. The emphasis is on the experimental side including various kinds of optical characterization, such as infrared spectroscopy, Raman measurements, luminescence characterization and perturbation spectroscopy, of confined donors and acceptors. Also the dynamical properties as derived from time resolved luminescence measurements are presented.

Linköping, Göteborg
June 2004

Per Olof Holtz
Qing Xiang Zhao

Contents

1 Introduction

The recent development of advanced growth techniques such as molecular beam epitaxy and metal-organic vapor phase epitaxy has made it possible to fabricate ultrathin semiconductor layers, quantum structures, with an accuracy down to atomic layer thickness. This means that dimensions smaller than the de Broglie wavelength of the electrons can be achieved. We have accordingly advanced from the classical physics to a situation with structures in the quantum limit. The driving force for the effort behind this development is the potentiality to modify the electronic properties of the quantum structures by the confinement [1–3].

As mentioned above, the door to this field has opened by the development of modern sophisticated growth techniques, by which dopant atoms can be introduced in a very controlled way both what concerns concentration as well as position within the quantum well or in the barrier. By varying the quantum well width, it is possible to follow the impurity properties from the pure bulk (3D) behavior to a quasi-2D one in the limit of narrow quantum wells. An interesting extension is to further confine the carriers in one-dimensional (1D) structures (quantum wires) or zero-dimensional (0D) structures (quantum boxes). There is currently a great effort to develop such low-dimensional structures of high quality. However, the development of these structures is still in an early stage for detailed optical studies of the impurities and is left outside the scope of this book.

2 Quantum Wells

The electrons in a 3D bulk crystal can be described by Bloch waves, which can propagate throughout the lattice in an unrestricted way. If this freedom is limited in one direction by introducing potential barriers, a two-dimensional nanostructure, a quantum well, is formed. In this way, the properties of electronic particles in the quantum well will be modified [1–3]. The electronic particles, described by Bloch waves, are spatially confined in two dimensions with a confinement energy, E, of the ground state for the particle with the effective mass, m^* (assuming infinite barriers)

$$E = \frac{\hbar^2 \pi^2}{2m^* L^2}, \tag{2.1}$$

where L is the well width. In this case, the barriers are assumed to be infinite. If we instead turn to a more realistic case with finite barriers, the electronic particles will experience a potential, which is described by

$$V(z) = \begin{cases} 0 & |z| < L_z/2 \\ V_0 & |z| > L_z/2. \end{cases} \tag{2.2}$$

Similarly, the effective masses of the electronic particles in the different regions are given by

$$m^*(z) = \begin{cases} m_A^* & |z| < L_z/2 \\ m_B^* & |z| > L_z/2. \end{cases} \tag{2.3}$$

m_A^* is the effective mass of the electronic particle in the well material A, while m_B^* is the corresponding effective mass in the barrier material B. In order to proceed with the theoretical treatment of the electron energy levels and wave functions of such a quantum well, we should solve the Schrödinger equation

$$-\frac{\hbar^2}{2} \frac{\partial}{\partial z} \left[\frac{1}{m^*(z)} \frac{\partial f_n(z)}{\partial z} \right] + V(z) f_n(z) = E_n f_n(z), \tag{2.4}$$

where E_n is the confinement energy of the electrons and f_n is the envelope wavefunction included in the electron wave function

$$\psi(r) = \sum \exp(ikr) u_k(r) f_n(z). \tag{2.5}$$

The envelope wavefunction, f_n, has an exponential tail into the barrier

$$f_n(z) = \begin{cases} A \exp(-kz) & z > L/2 \\ B_1 \sin(\gamma z) + B_2 \cos(\gamma z) & |z| \leq L/2 \\ C \exp(kz) & z < -L/2 \end{cases} \qquad (2.6)$$

in contrast to the case with an infinite barrier. The boundary conditions require the continuity of the wave function as well as its derivative at the interfaces. The latter condition can in a more general respect be described in terms of the continuity of the particle current, which allows a change in the effective mass according to

$$\frac{1}{m_1^*} \frac{\mathrm{d}f_1}{\mathrm{d}z} = \frac{1}{m_2^*} \frac{\mathrm{d}f_2}{\mathrm{d}z} . \qquad (2.7)$$

The solutions of the Schrödinger equation above will provide us with energy eigenvalues. The number of bound states is given by

$$1 + \mathrm{Int}\left[\left(\frac{2m^* V_0 a^2}{\pi^2 \hbar^2}\right)^2\right], \qquad (2.8)$$

i.e., there will be at least one bound state in any quantum well. The bound state at lowest energy will be of even parity, the next one of odd parity and so on. There are applicable selection rules governing the transitions allowed in this system [4].

3 Impurities in Bulk

The possibility to incorporate impurities in the semiconductor crystal to change its optical and electrical properties is of paramount importance for the applications of semiconductors. The incorporation of an impurity to the semiconductor crystal corresponds to an effective addition of a charge carrier and a charged impurity ion to the system. The impurities will give rise to localized states in the forbidden energy gap. Based on the energy separation between the localized state and the band edge, the impurities are referred to as deep or shallow states.

3.1 Effective Mass Theory

If the binding energy of an electron (a hole) to a donor (an acceptor) is relatively small, the impurity energy levels can be estimated by treating the impurity potential as a perturbation of the periodic potential of an otherwise perfect lattice. Due to the Coulombic attraction, the electron (hole) will move in an orbital motion around the donor (acceptor). The orbiting electron (hole) will experience the Coulomb potential screened by the dielectric medium of the host valence and core electrons. The foundation of the effective mass theory is the analogy with an isolated hydrogen nucleus. The donor (acceptor) binding energy is derived from a simple Bohr model by substituting a reduced effective electron (hole) mass and a Coulomb potential screened by the dielectric of the host material. In the effective mass theory, the solutions for the hydrogen atom with appropriate substitutions of the effective mass and the semiconductor permittivity are used. The result is a series of hydrogenic impurity levels given by

$$E_B^n = 13.6 \frac{Z \, m^*}{n^2 \, \epsilon^2 \, m} \text{ (eV)}, \tag{3.1}$$

where 13.6 eV is the hydrogen ionization energy and Z corresponds to the number of protons in the nucleus, e.g., for the hydrogen atom, $Z = 1$. For materials with a nearly spherical conduction band like GaAs, the results derived from this hydrogenic model are in good agreement with the experimental results for shallow impurities. But also for deep impurities with a ground state bound by an energy considerably exceeding the effective mass-like value, the excited states associated with this deep impurity can still be effective mass-like. For shallow impurities, like donors and acceptors to be discussed in the following, the binding energy E_B and the effective three dimensional Bohr radius a_0 of the ground state can be estimated by the following formula,

$$E_B = 13.6 \, \frac{m^*}{\epsilon^2 \, m} \; (\text{eV})$$

$$a_0 = 0.53 \times \epsilon \times \frac{m}{m^*} \; (\text{Å}) \,. \tag{3.2}$$

The above formulas are derived by assuming that the hydrogen-like impurity is correlated to a nondegenerated and isotropic band, which is true for most of shallow donor cases.

3.2 Donors

The donor binding energy can be estimated with a high measure of precision by (3.2) due to the very small and isotropic electron effective mass. This also means that the envelope motion of the donor electron can be accurately described in terms of the scaled hydrogenic wave functions.

In GaAs, $\epsilon = 12.85$ and $m^* = 0.067 \, m$ for electrons, so the Bohr radius and the binding energy of the ground 1S donor state can be estimated to $a_0 \approx 100 \,$Å and $E_B \approx 5.5 \,$meV, respectively. The smaller binding energy of donors results in a very small energy shifts between different donors, sometimes just fractions of a meV. This fact by itself makes an identification of the donors more demanding than for acceptors, what concerns the energy resolution for the technique employed to identify the donor. Far infrared photoconductivity is a high resolution method in which the 1s–2p transition of the donor, is the usually monitored transition. If these measurements are performed in the presence of a magnetic field, the transitions to different 2p states can be identified. In this way, donors have been identified in pure epitaxial GaAs films [5].

Another spectroscopic technique often used for identification of donors is the two-electron satellites observed by resonant excitation of the donor bound exciton. Usually, the electron of the donor, which binds the exciton is normally left in its ground state after the recombination, but there is a small but non-zero probability that this donor electron is left in an excited state after the recombination instead. This will give rise to a satellite, which is red-shifted relatively the principal bound exciton with an energy corresponding to the excitation energy required to excite the donor electron into its excited state. With this technique, the energy separation between the principal bound exciton and the satellite can be determined with a high accuracy. Accordingly, this method will allow us to spectroscopically resolve and identify donors, which are very close in energy. A nice illustration on what can be achieved with this technique was demonstrated by T.D. Harris et al. [6], who were able to spectroscopically identify the shallow donors in GaAs. The same technique has later on been applied also on excitons bound at confined donors, as will be further expounded below (in Sect. 4.2.4). Furthermore, by detecting a two-electron satellite in a luminescence excitation spectrum, T.D. Harris et al. [6] opened the possibility to investigate the electronic structure of the bound exciton. The resolution improvement observed was found to be larger in the luminescence excitation spectrum than in the resonant photoluminescence measurement due to elimination of interfering emission.

A further spectral resolution improvement was achieved by performing the luminescence measurements in a Fourier Transform spectrometer with a resolution of a few μeV [7]. Several excited states, both s-like and p-like, of the donor bound exciton were monitored. In high resolution selective photo-luminescence and photoluminescence excitation experiments monitoring the two-electron satellites, the extremely small chemical shifts of the donors in GaAs were resolved. The excited state spectroscopy was extended to include also the magnetic field dependence. Any orbital state of the donor, the electron can have either the spin projection $m_\mathrm{s} = +1/2$ or $m_\mathrm{s} = -1/2$ on the magnetic field. The g-factor for electrons in the conduction band in bulk GaAs is $g = -0.44$ [8]. Accordingly, the $m_\mathrm{s} = +1/2$ state of the donor has a lower energy than the $m_\mathrm{s} = -1/2$ state. The g-factor for the donor could be derived from the transition energy from the 1s ground state to the 2p-excited state with either the spin projection $m_\mathrm{s} = +1/2$ state or $m_\mathrm{s} = -1/2$. Consequently, the energy separation between these components correspond to the spin-flip energy $\Delta E = \mu_\mathrm{B} g_\mathrm{e} B$, where μ_B is the Bohr magneton. The experimentally determined value on g_e was compared and found to be in good agreement with theoretical predictions based on appropriately scaled calculations for the hydrogen atom as derived from the $\boldsymbol{k} \cdot \boldsymbol{p}$ theory.

3.3 Acceptors

While the donor states close to the non-degenerate conduction band can normally be described by a single band effective mass approximation, the corresponding description of acceptor states close to the valence band edge is more complex due to the degenerate valence band (six-fold degenerate at Γ). A pioneer work to calculate such states was performed by E.O. Kane [9] in the $\boldsymbol{k} \cdot \boldsymbol{p}$ theory, in which a multi-band effective mass approximation was employed.

Near the Γ point in the Brillouin zone, the upper valence band consists of three p-like states and is consequently sixfold degenerate, if the spin is taken into account. These states can be characterized by the total angular momentum quantum number, $J = 1/2$ and $J = 3/2$. Including also the spin-orbit interaction, these states are split, leaving the $J = 3/2$ band lowest in energy. The corresponding acceptor ground state is denoted $1S_{3/2}(\Gamma_8)$. This band is thus fourfold degenerate at $\boldsymbol{k} = 0$ in the bulk case, but splits into two Kramers doublets, the heavy hole (hh) and light hole (lh) states, for $\boldsymbol{k} \neq 0$ (see Fig. 3.1).

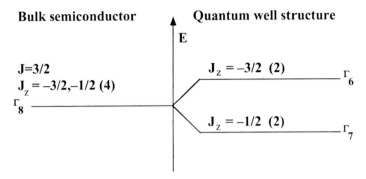

Fig. 3.1. Schematic figure of the acceptor ground state which in bulk is denoted $1S_{3/2}(\Gamma_8)$ with a 4-fold degeneracy. In the case of a quantum well, this 4-fold degenerate band is split into two 2-fold degenerate hole bands

3.4 Isovalent Centers

Although isovalent centers are electrically inactive, they can still bind excitons under certain conditions. There are two main factors, which affect the ability to bind excitons: The electronegativity of the isovalent center in comparison with the host atom it replaces, i.e. the short range potential can bind an electron for the case of a greater electronegativity of the isovalent center and a hole, if the electronegativity is smaller than the host atom it replaces. Once this primary particle is bound, the long range Coulomb potential of the primary particle can bind a secondary particle of opposite sign in an effective mass-like state. The second factor of importance for the ability of isovalent centers to bind excitons, is the strain field due to the different size of the isovalent center relatively the host atom. Similarly to the electronegativity factor, the strain field can be attractive to either electrons or holes dependent on whether the extension of the isovalent center is larger than the host atom or vice versa.

The isovalent centers have been extensively studied both from an experimental and a theoretical point of view. The experimental work refers to different isovalent center systems in bulk material, like GaP [10–12], ZnTe [13,14], Si [15] and GaAs [16]. On the theoretical side, no first-principle theory exists, but the properties of the isovalent centers are by now well understood based on the experimental and modelling work. The most well known model is the Hopfield–Thomas–Lynch (HTL) model [17]. According to this model, the primary particle of the exciton bound at the isovalent center is attracted by the short range impurity and/or defect potential. The secondary particle is then attracted by the Coulomb potential of the primary particle. Based on this approach, we can distinguish between electron and hole attractive isovalent centers according to whether the primary particle is an electron or hole.

The perturbation Hamiltonian for an isovalent center is based on a hole state with either an angular momentum of $J = 1/2$ or $J = 3/2$. The Hamilto-

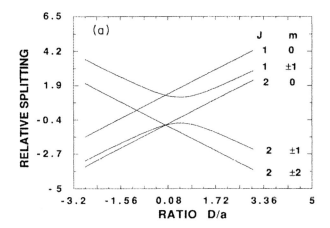

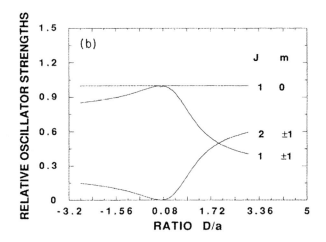

Fig. 3.2. (a) The predicted dependence of the exciton energy-level splitting on the ratio D/a, where the D parameter represents the strength of the crystal field potential and a represents the strength of the electron-hole interaction. (b) The relative oscillator strengths of the electric-dipole-allowed transition exciton states vs. the ratio D/a. In this figure, J represents the total angular momentum of the bound exciton, which is formed by a hole with a 3/2 angular momentum and an electron with a 1/2 angular momentum

nian for such a center, involves the spin-orbit interaction as well as the crystal field potential. The spin-orbit interaction is measured in terms of the spin-orbit splitting constant, Δ. The crystal field potential affects only the orbital momentum of the holes. For the case of a strong crystal potential, expressed by a large D parameter, with a low symmetry, the orbital angular momentum degeneracy of the hole is broken [18]. The wave functions of the bound hole states are spin-like. This means that the orbital angular momentum of the bound hole is quenched and accordingly $J = 1/2$. The spin-like holes and electrons interact to give singlet-triplet states. The effective Hamiltonian of the bound exciton can be written [19]

$$H_{\text{exciton}} = -A\,\boldsymbol{S} \cdot \boldsymbol{J} - D\left[J_z^2 - \frac{J(J+1)}{3}\right],\qquad(3.3)$$

where J and S is the angular momentum of a bound hole and bound electron, respectively. The parameters a and D represent the strength of the electron-hole interaction and the crystal field potential, respectively. The adequate oscillator strength for the associated bound exciton has been calculated by Q.X. Zhao and T. Westgaard [19] (see Fig. 3.2).

In the other extreme case, with a small D parameter, i.e. the spin-orbit interaction dominates over the crystal field, the bound hole can be treated as a particle with the angular momentum left unchanged ($J = 3/2$). The coupling between the resulting energy levels is negligible. The perturbation Hamiltonian for this case has been discussed in detail, also in the presence of a magnetic field, by e.g., Q.X. Zhao and B. Monemar [20]. For the intermediate case, with moderately strong crystal field, the mixing between the states has to be taken into account and it is consequently not possible to assign a total angular momentum of $J = 1/2$ or $3/2$ of the bound hole states.

4 Confined Neutral Donor States

The early calculations on the impurity states in quantum wells were based on the one-band effective mass approximation (EMA), which earlier has been successfully applied on the corresponding shallow impurity states in 3D bulk material. In the original theoretical work for the confined impurity states, Bastard performed calculations on the electronic structure of the impurities in quantum wells assuming a hydrogenic impurity confined in a quantum well with infinite barrier heights [21]. In this approximation, the electronic wave function vanishes in the barrier and the wave function of an impurity at the interface must be a p-like state, whereas the corresponding wave function for an on-center impurity is a s-like state. For the case of an impurity at the interface and in the center in the wide quantum well limit, a binding energy of $R^*_{3D}/4$ and R^*_{3D}, respectively, is predicted (with R^*_{3D} being the effective Rydberg for a hydrogenic impurity in bulk). Also the calculated binding energies for an on-center impurity was found to increase monotonically from the bulk value, R^*_{3D} to $R^*_{2D} = 4R^*_{3D}$ at vanishing quantum well width, L_z (see Fig. 4.1).

In more realistic calculations, finite barrier heights were used, which allows the wave function to penetrate into the barrier [22–25]. When $L_z \ll a^*$ [26], the confinement effect from the quantum well potential dominates the Coulomb potential from the impurity and the envelope function is similar to the intrinsic one (the envelope for a quantum well without any impurity). In the intermediate case, $L_z \sim a^*$, the effects due to confinement and Coulomb attraction are comparable. The derived impurity binding energy exhibits the characteristic shape shown in Fig. 4.2. In this case, the derived binding energy reaches a maximum for a non-zero value on the quantum well width (between 20 and 50 Å) to be back at the value corresponding to the barrier material in the limit $L_z \to 0$.

4.1 Theoretical Aspects

In order to calculate the shallow donor states in quantum wells, the effective mass theory is applied on the donors, usually by means of a variational method of solution. The pioneer work in this field was performed by G. Bastard [21], who considered the donor states associated with the first subband. Furthermore, he assumed infinite barriers, i.e., the donor wavefunctions do

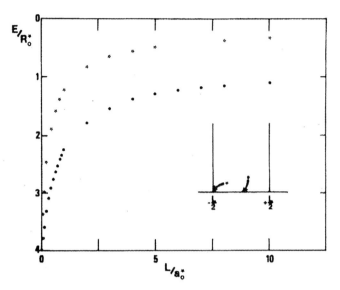

Fig. 4.1. The reduced binding energy of a donor confined in a quantum well structure versus reduced layer thickness for a donor located at the center (*dots*) or at the boundary (*stars*) of the well. As the well width decreases asymptotically towards 0 (the true 2D case), the binding energy of the on-center donor increases monotonically from the 3D value (E_b) in the bulk limit ($E_b = 5.8\,\text{meV}$ in GaAs) to $4\,E_b$ as the well width vanishing. (From [21])

not penetrate into the barrier. The justification for this argument was found in the fact that the donor binding energy is very small in comparison with the conduction band offset. G. Bastard performed calculations of the donor binding energy as a function of the well width as well as the donor position within the quantum well. Later on, the theoretical predictions were improved by introducing more realistic barriers with a finite height. Such calculations were subsequently reported by C. Mailhiot et al. [27,28] and R.L Greene and K.K. Bajaj [24].

4.1.1 The Effective Mass Approximation

On the concept of the effective mass approximation applied on an electron bound at a donor in a quantum well, the effective mass Hamiltonian is given by [25]

$$H = \frac{\hbar^2 k^2}{2m^*} + H^{\text{QW}} + H^{\text{C}}. \tag{4.1}$$

Here H^{QW} represents the confinement potential and H^{C} is the Coulomb potential for a point charge in a certain dielectrics, separated from the surrounding different dielectrics by two infinite planes.

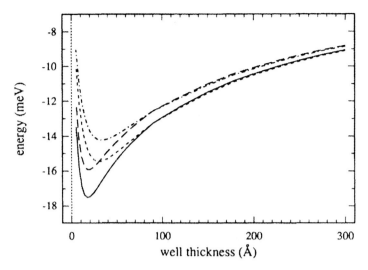

Fig. 4.2. The dependence of the energy of the donor ground state, 1s, on the well thickness L, for an on-center impurity and for $x = 0.3$ (*solid line*) in a GaAs/Al$_x$Ga$_{1-x}$As quantum well. The results represented by the *solid line* were obtained by using the proper values of the dielectric constants and effective masses in the well and barriers materials; *dashed line* shows the analogous results when approximations in the parameters ϵ_2, m_2^* are used; *long-dashed line* is obtained for $\epsilon_2 = \epsilon_1$; *short-dashed line* is derived for $m_2^* = m_1^*$; *dot-dashed line* for both approximations. The energy scale shown in the figure refers to the bottom of the first subband in the quantum well. These predictions clearly illustrate that the derived binding energy reaches a maximum for a non-zero value on the well width (20–50 Å) to asymptotically approach the value corresponding to the barrier material in the limit $L_z \to 0$. (From [26])

Also, the dependence of the donor position within the well for different well width and various alloy composition was investigated. The binding energy as a function of well width for the two limiting cases; an on-center donor and an on-edge donor, exhibit a qualitatively similar shape. In the thin well limit the donor binding energies are very similar for the on-center and on-edge donor, while the edge donor is significantly less tightly bound in the thick well limit. C. Mailhiot et al. [27, 28] explains this in terms of an increasingly number of Block states from the AlGaAs conduction band edge required to construct the ground state envelope function as the donor is moved away from the center to approach the edge. As the on-edge donor ground state envelope function contains more of these higher states, the on-edge donor becomes more shallow relatively the on-center donor. Another fact to take into account is the repulsive barrier potential, which tends to increase the spatial separation between the electronic charge distribution and the ionized donor, which in turn results in a reduced Coulomb attraction.

Also the energies of excited s-like donor states have been theoretically predicted in some approaches [26, 29]. Stopa and DasSarma solved numerically the Schrödinger's equation and determined the 1s–2s transition energy of donors in the center of the quantum wells [29]. Fraizzioli et al. introduced different effective masses and dielectric constants in the well and in the barrier into their calculations based on a variational method [26]. This correction was found to be quite sizeable for in particular large x values (high Al concentration) and small well widths.

By varying the well width of a quantum well, the behavior of the carriers can be followed from the three dimensional (3D) case to the quasi-2D one in the limit of very narrow wells. In particular, the extrinsic properties due to shallow impurities are considerably influenced by the spatial confinement. The donor states are simpler to treat due to the fact that they are associated with the non-degenerate conduction band. In the simplest approach when the donors are located in the center region of well layer, the donor states are considered as a hydrogenic impurity confined in a well with unpenetrable barriers. As the well width decreases, the binding energy of an on-center donor increases monotonically from the 3D value (E_b) in the bulk limit ($E_b = 5.8\,\text{meV}$ in GaAs) to 4 (E_b) as the well width asymptotically approaches 0 (in the true 2D case) (see Fig. 4.1). In more realistic calculations, the barriers were treated as finite [30]. For the excited states such as p_0 and $p_\pm$, the theoretical treatments have been discussed in the following way [31],

$$H^{\text{QW}}(z) = \left\{ \begin{array}{lll} 0 & |z| < & L_z/2 \\ V_0 & |z| > & L_z/2 . \end{array} \right. \tag{4.2}$$

For an electron bound at a donor in such a well, the Hamiltonian in the effective-mass theory is given by [21]

$$H = \frac{-1}{m^*}\nabla^2 - \frac{1}{r} + H^{\text{QW}}(z) . \tag{4.3}$$

In many cases, a variational wave function of the type

$$\psi(\rho, z, \phi) = f(z)\, G(\rho, z, \phi) \tag{4.4}$$

is used, where the function $f(z)$ is a solution for the one-dimensional square well and $G(\rho, z, \phi)$ corresponds to the trial wave function for the electron relatively the donor. In order to calculate the 2p-like states of the donor, the unnormalized function $f(z)$ is the analytical solutions for the first excited state in a square well with an infinite barrier height, i.e.

$$f(z) = \left\{ \begin{array}{ll} -A\exp(\kappa z) & z < -L/2 \\ \sin(kz) & |z| \le L/2 \\ A\exp(-\kappa z) & z > L/2 . \end{array} \right. \tag{4.5}$$

L is the well thicknes, the parameter k is determined by the first excited state energy. A and κ are fixed by matching conditions at the interface of the

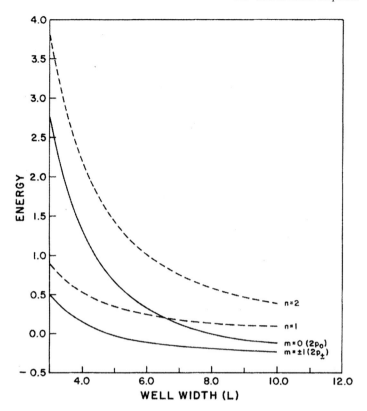

Fig. 4.3. The theoretically predicted binding energies of a confined donor for the $2p_\pm$ and $2p_0$ like states versus the thickness of the GaAs well (L) (*solid lines*). The corresponding energies for the first and second subband edges are also given versus the GaAs well thickness. Note the peculiar behaviour of the $2p_0$ state as demonstrated by Greene and Bajaj [31]: As the binding energy is reduced with decreasing well width, the binding energy becomes in fact zero for a certain well width (about 650 Å; from [31])

quantum well. In this case, the binding energy of a donor should come back to the 3D value of the barrier material, as the well width asymptotically approaches zero. The result of such calculations performed for three odd-parity states, labeled by their quantum number m, $2p_\pm$ and $2p_0$, as a function of well width, L_z, is shown in Fig. 4.3 [31]. In the same figure, the corresponding first and second subband edges are also given. A certain peculiarity of the $2p_0$ state was demonstrated by Greene and Bajaj [31], namely that its binding energy is reduced as the well width decreases. For a certain well width (about 650 Å), the binding energy in fact becomes zero. This effect can be realized from Fig. 4.3, since the $2p_0$ state is associated with the second subband and remains bound with respect to that and not the lowest subband as the well width is reduced.

4.1.2 The Presence of an External Field

An external field can be an electric field, a magnetic field or a uniaxal pressure. Since the conduction band has only spin degeneracy, an uniaxal pressure (or built-in strain in the lattice-mismatched quantum well system) has no influence on the binding energy, except for the donor states corresponding to different conduction band minima shift from each others if the conduction minimum is not located in the Γ point, e.g., for the case of Si. Therefore, the presence of electric and magnetic fields will be discussed here. When an electric field is applied along the growth axis, a spatial shift of the wave function is obtained. This results in turn in the Stark shift, a spectral red-shift. The first calculations were performed by applying an electric field on an infinite quantum well [32] and later on for the finite well case by Ahn and Chuang [33]. The effect of the electric field is introduced into the Hamiltonian for the electron in an effective-mass approximation by a term $\phi(z) = e\,z\,E\,a_{\mathrm{B}}^*/(\hbar^2/m^*\,a_{\mathrm{B}}^{*2})$, where E is the electric field and a_{B}^* is the effective Bohr radius. B. Sunder [34] variationally calculated the binding energy of a hydrogenic impurity confined in an infinite quantum well by minimizing the expectation value of the Hamiltonian for the electron. The energy is found to monotonically decrease with increasing ϕ, i.e., with the applied electric field, but the shift is different for the ground state and the excited state, but also different for different excited states with the applied field. This fact is illustrated in Fig. 4.4 for the case of some different quantized levels compared with the ground state. For the $n = 1$ ground state, the binding energy will go to zero for a certain field (67 kV/cm for a well width corresponding to the effective Bohr radius). For excited states, the binding energy will go to zero for a higher field. It was also shown that when the well width is increased, the binding energy will go to zero at a lower field.

Experimentally, the effect of an applied electric field on the quantum well states has been estimated by electroreflectance [35, 36]. A monotonic decrease of the energies for the ground state and the first excited states with an increasing field was observed. However, a different behavior is found for the higher quantized levels, when an electric field is applied. In fact, even a blue-shift has been reported for higher quantized levels with a field applied [37]. Also, the effect of an applied electric field on the excited well states as deduced from photoluminescence measurements has been reported [38].

The effect of an electric field on the electronic structure has also been investigated by means of infrared spectroscopy. The transitions from the ground state to excited states of hydrogenic donors were monitored in the presence of a magnetic field by Yoo et al. [39]. The transition energy decreases dramatically as the electric field increases. The predominant 1s–2p transition is red-shifted by as much as 75%, when an electric field of 2.3×10^4 V/cm is applied over the structure. It is assumed that the major voltage drop occurs across the intrinisic region involving the quantum wells of the p–i–n structure used. The large red-shift observed (up to 75%), opens possibilities for different

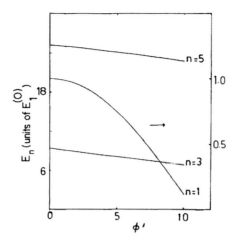

Fig. 4.4. The energy of an electron confined in an infinite quantum well as a function of the electric field parameter ϕ^1 given by $\phi^1 = e\,E\,l/E_1^0$, where E and l are the applied electric field and the well width, respectively. For the $n = 1$ ground state, the binding energy will asymptotically approach zero for a certain field ($67\,\text{kV/cm}$ for a well width corresponding to the effective Bohr radius), while the binding energy will reach zero for higher fields for the excited states. (From [34])

kinds of application such as infrared detectors. A photoconductive response originating from confined donors have been reported [40, 41]. The use of an externally applied electric field could make it possible to control the frequency response of the donor photoconductivity in quantum well structures.

An applied magnetic field along the growth direction of the quantum well structure will break the time-inverse asymmetry, resulting in a splitting of each spin degenerate donor state. A number of theoretical calculations have been performed to study the magnetic field dependence of donor states [30, 42–45]. The models used are either a variational approach or a perturbation approach. The binding energies of the ground (s-like) and excited (2p$_+$- or 2p$_-$-like) states of a donor confined in the quantum well structures have been calculated as a function of magnetic fields and well widths. For a given value of the magnetic field, the binding energies are found to be larger than their values in a zero magnetic field, and the 2p$_+$ state shows a larger change in comparison with other states with the magnetic field strength. Similar calculations have been performed for donors confined in a coupled double quantum well structure [46]. When an applied magnetic field is perpendicular to the growth direction, the calculation is more complex. Calculations on the transition probability between the ground state and the excited states yield the theoretically most probable transition energy for a given magnetic field [47].

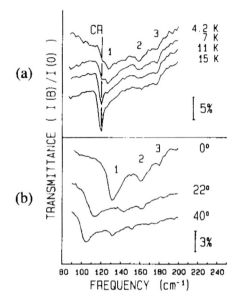

Fig. 4.5. (a) The temperature dependence for the infrared transitions between the 1s-like donor ground and the excited $2p_\pm$-like states measured by means of far-infrared transmission measurements in the presence of a field of 9 T. (b) The angular dependence of the absorption spectra in barrier and center donor doped quantum wells measured at a field of 9 T. The angle notation corresponds to the angle between the magnetic field and a normal to the sample. The transmission data correspond to the ratio of the detected signal at the indicated magnetic field to the detected signal at zero magnetic field. (From [49])

4.2 Experimental Aspects

In the following a survey of the most important optical techniques to characterize the confined impurity states will be presented. The major part involves stationary characterization, but also the dynamical properties of the impurities studied by means of time resolved spectroscopy will be highlighted in Sect. 4.2.6.

4.2.1 Infrared Measurements

In infrared transmission measurements, transitions between the 1s-like donor ground and the excited $2p_\pm$-like states are predominant, since transitions between states with different parity are allowed according to selection rules applicable for these transitions. Due to the superior sensitivity of the Fourier Transform Interferometric spectroscopy, this technique is an often utilized technique for optical measurements in the infrared range. This is a versatile instrument used for various kinds of spectroscopy such as absorp-

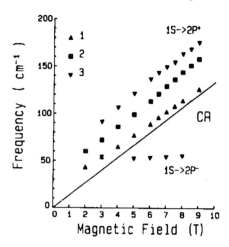

Fig. 4.6. The frequency dependence of some different transitions between the 1s-ground and the excited $2p_\pm$-like states measured by Fourier Transform Infrared spectroscopy in the transmission mode at different magnetic fields. The lines indicated in the figure correspond to: (1) the 1s–2p ($m = +1$) transition for donors at the center of the wells; (2) the transition from the ground state to the second excited state with $m = +1$ for barrier impurities; (3) the transition from the ground state to the first excited state with $m = +1$ for barrier impurities. The solid line refers to the cyclotron resonance results. (From [49])

tion, transmission, reflectivity and photoconductivity, but also photoluminescence and photoluminescence excitation spectroscopy. In a pioneering work in this field performed by N.C. Jarosik et al. [48], far-infrared magnetospectroscopy was carried out on confined shallow donor states in the center of the GaAs/AlGaAs wells of varying thickness. Later on, Reeder et al. [49] further studied the transitions of the shallow donor states by far-infrared transmission measurements at different temperatures and magnetic fields (see Fig. 4.5). In the confining potential of a quantum well, the degeneracy of the 2p states is removed, leaving the $2p_0$ state at highest energy, while the $2p_\pm$ remains degenerate at zero field. This is nicely illustrated in a transmission measurement performed by Fourier Transform Infrared spectroscopy [49] (see Fig. 4.6). In the presence of a magnetic field, also the degeneracy of the $2p_\pm$ states is removed with the $2p_+$ state left at highest energy. This fact was seemly demonstrated by N.C. Jarosik et al. [48] in their magneto-spectroscopy measurements, in which the splitting between the 1s–$2p_+$ and the 1s–$2p_-$ transitions of the confined donor were monitored. As expected, the transition energies occur at a higher energy relatively the corresponding bulk donor energies in the presence of a magnetic field. Another interesting observation in these magnetic field dependent spectra, is that the blueshift caused by the confinement is slightly smaller at higher magnetic fields. This can be explained in terms of an increasing localization

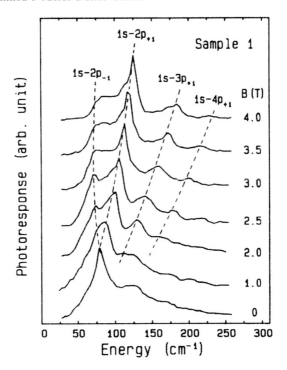

Fig. 4.7. Photoresponse spectra from photothermal ionization spectroscopy (PTIS) measurements for the confined donor transitions measured with a far infrared light propagating normal to the sample surface in the presence of a varying magnetic field parallel with the infrared light at 4.2 K. (From [50])

of the donor wave function at increasing magnetic fields. This means that the quantum confinement effect becomes decreasingly important at higher magnetic fields. When the sample is rotated relatively the magnetic field, it is shown that all three lines approximately scale with the component of the magnetic field, which is along the growth direction. This fact tells us that these three lines are confinement related.

Later on, well-defined transitions to excited states of the confined donor have been detected in photo-thermal ionization spectroscopy (PTIS) [50]. In the first step, donor electrons are optically excited to a p-like excited state. In a second step, the electrons are thermally ionized from the excited state into the conduction band, which in turn gives rise to a photocurrent. These measurements were performed by means of Fourier Transform spectroscopy with a far infrared light propagating normal to the sample surface in the presence of a magnetic field parallel with the infrared light. Photoconductivity spectra were measured at different magnetic fields as shown in Fig. 4.7. At zero field, the predominant feature in this photoconductivity spectrum is the 1s–2p transition, but in the presence of a magnetic field, this line is split into two components; 1s–2p$_-$ and 1s–2p$_+$. Also well-defined transitions to higher p-like states, 1s–np, are nicely demonstrated in these photoconductivity spectra. By inspecting the magnetic field dependence of each transition, it is apparent that the 1s–3p$_{+1}$ transition has a slope of about $2\hbar\omega_c$ and 1s–4p$_{+1}$ transition has a slope of about $3\hbar\omega_c$ in the high field region. Therefore the experimental results clearly indicate that states 3p$_{+1}$ and 4p$_{+1}$ move with the N=2 and N=3 electron Landau levels in the high magnetic field region. As the 1s–2p$_+$ transition of the confined donor is tuned towards significantly larger energies at high magnetic fields, and approaches the GaAs optical phonon energies.

4.2.2 Raman Measurements

In Raman scattering experiments, the selection rules allow transitions to excited states with the same parity as the ground state. Shanabrook et al. [51] have presented results from such Raman experiments, in which the excitation energy was kept resonant with the transition between the spin-orbit-split valence band and the lowest conduction band of the quantum well, a broad peak ($\Delta_{FWHM} > 5$ meV) interpreted as the 1s–2s Raman transition of the donor was observed. The extensive investigations of donors by Raman spectroscopy have essentially been performed on II–VI quantum well structures [52–57]. In the dilute-magnetic semiconductor CdTe/Cd$_{1-x}$Mn$_x$Te quantum wells structures, the spin-flip Raman scattering of electrons bound to donors has been studied for various thicknesses and Mn-concentrations. For structures with $x < 0.1$, two spin-flip Raman bands are observed, which have been assigned to electrons located in the quantum wells and bound to donors located either in the quantum wells themselves or in the barriers of the structure [53]. The experimentally observed spin-flip Raman shifts for different specimens have been compared with the calculated results, and consequently by the comparison the conduction-band offset and the ratio of this quantity to the total band-gap difference between wells and barriers can be deduced. Furthermore, the spin-flip Raman scattering from donor bound electrons in Cd$_{1-x}$Mn$_x$Te/Cd$_{1-y}$Mg$_y$Te single quantum wells have also been investigated [55]. An enhancement of the scattering intensity of the bound magnetic polaron was found for narrow quantum wells. The Raman shift was found to strongly depend on the thickness of the quantum well and the applied magnetic field. The high-field Raman shifts decrease with decreasing well thickness, as shown in Fig. 4.8. The situation is more complex in the low-field limit where the bound magnetic polaron (BMP) effect has to be considered.

4.2.3 Luminescence Measurements

Similarly to bulk, the luminescence associated with extrinsic, i.e., impurity related, as well as intrinsic processes are observed in quantum wells. However, while the extrinsic processes are usually predominant in the luminescence

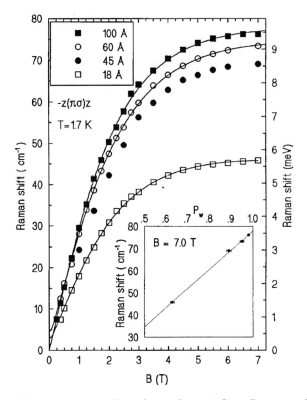

Fig. 4.8. The experimentally observed spin-flip Raman shifts as a function of the magnetic field for donor-bound electrons confined in $Cd_{0.97}Mn_{0.03}Te/Cd_{0.76}Mg_{0.24}Te$ single quantum wells with four different thicknesses. The high-field Raman shifts decrease with decreasing well thickness. The insert shows saturation value of the Zeeman splitting versus the calculated probability of finding the electron of the first subband within the potential well. (From [55])

measurements of bulk samples, also intentionally undoped, the situation is significantly different in quantum wells. This is illustrated in Fig. 4.9, in which the donor bound exciton (DBE) is not comparable in intensity with the free exciton (FE) until the doping level is as high as $5 \times 10^{16} \ cm^{-3}$. In this case, the intrinsic recombination processes dominate the luminescence spectra until fairly high impurity concentrations. The radiative recombination decay times are considerably shorter in a quantum well than in bulk [58], which in turn results in a reduced probability for capture of the charge carriers at impurities relatively radiative recombination. Furthermore, as the well width decreases, the electron-hole wavefunction overlap increases, resulting in a reduced observed decay time [59]. The photoluminescence associated with Si donors confined in quantum wells was first reported by B.V. Shanabrook

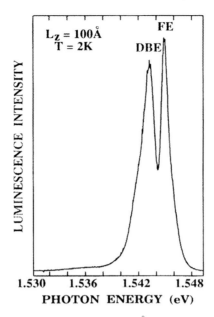

Fig. 4.9. Photoluminescence spectrum of a 100 Å wide quantum well, doped with Si to a level of $5 \times 10^{16}\,\mathrm{cm}^{-3}$ measured at 1.7 K with an excitation energy of 1.569 eV. Only the free exciton and the donor bound exciton are observed in the spectrum

and J. Comas [60, 61]. A novel emission band at an energy slightly below the heavy hole exciton was reported and interpreted as the recombination between a bound Si electron and a hole from the valence band, i.e., a free-to-bound recombination. This observation made it possible to evaluate the Si donor binding energy as the energy separation between the band edge and the free-to-bound transition energy.

Recombination Processes

In bulk semiconductors, the bound exciton was first predicted by M. Lampert in 1958 [62]. He made the analogy between the excitons weakly bound at acceptors or donors and similar atomic and molecular systems. The first experimental evidence for a bound exciton was reported by J.R. Haynes [63] by recognition of some sharp lines in addition to the free exciton appearing in Si. These novel lines appeared at a slightly lower energy than the free exciton as the Si was doped with As donors. The small energy difference between the free and bound exciton, i.e., the limited binding energy for the bound exciton directly implies that the exciton is bound at a *neutral* host atom. Later on, J.J. Hopfield [64] claimed that *neutral* acceptors/donors should always exhibit a finite binding energy for the bound exciton, while *ionized* impurities should only bind for special circumstances (regarding for instance the effective mass ratio m_e^*/m_e [65]). Also, one is most often dealing with effective mass-like impurities, which bind excitons. In general, complex centers with large binding energies, such as native defects, interstitials and vacancies do not bind excitons, although exceptions occur.

Most of the existing information on the bound excitons has been obtained from optical spectroscopy. In parallel with an improved knowledge on the exciton system, considerable information has also been attained on the acceptors or donors, which bind the excitons, as will be illustrated several times later on in this book. In fact, the optical spectra often contain very detailed information on e.g. the electronic structure of the host impurity. However, this kind of fine structure becomes obscured by thermal broadening effects already at very low temperatures. Accordingly, the detailed and informative optical spectra of the bound exciton systems have to be recorded at very low temperatures. D.G. Thomas and J.J. Hopfield [66] made a pioneer work in the field of the fundamental properties of the bound excitons. They investigated for instance the transition oscillator strength, the electron-hole j–j coupling, the phonon coupling and the qualitative estimate of the binding energies of the bound excitons. For the case of an exciton bound at the neutral donor, the two electrons are paired in the exciton ground state. The resulting energy level structure from the j–j coupling between the two electrons and the single hole is a simple $J = 3/2$ state (see Fig. 4.10). In luminescence, the transition from this initial $J = 3/2$ exciton state to the $J = 1/2$ neutral donor ground state gives rise to a single no-phonon line. The donor bound excitons in GaAs all appear within a very limited energy range (about 0.1 meV), which implies differences in the donor ionization energies of the order of 1 meV. These small energy separations between the bound excitons result in difficulties to resolve the excitons and the subsequent identification of the corresponding donor binding the exciton. In this respect, the observation of the so-called two-electron transitions has been essential, since the energy splitting between the satellites are approximately an order of magnitude larger than the corresponding energy displacement between the donor bound excitons. The observation of satellites provides accordingly strong support for the interpretation of the suggested identification of the donor bound excitons (see Sect. 4.2.4 below). Furthermore, the existence of these satellites also opens the possibility to monitor a selected principal bound exciton line in photoluminescence excitation spectra by detecting a specific satellite associated with a certain donor [67].

In the presence of a magnetic field, the Zeeman spectrum of the donor bound exciton recombination is determined by the single hole Hamiltonian as the initial state, given by [68,69]

$$E^h = E_0^h - 2\mu_B \left[\kappa \, \mathbf{J} \cdot \mathbf{H} + q \left(J_x^3 H_x + J_y^3 H_y + J_z^3 H_z \right) \right] + \frac{1}{\mu_B} \left(\frac{e \, a_0}{2c} \right)^2$$
$$\times \left\{ C_1 H^2 + C_2 H \left(\mathbf{J} \cdot \mathbf{H} \right)^2 + \frac{1}{2} C_3 \left[H_x H_y (J_x J_y + J_y J_x) + \text{c.p.} \right] \right\} \quad (4.6)$$

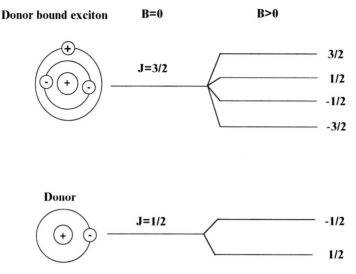

Fig. 4.10. Schematic illustration of the confined donor and its bound exciton with and without the presence of an external magnetic field

and the $J = 1/2$ neutral donor ground electron state as the final state. Fitting of the experimental results from the Zeeman spectrum to the Hamiltonian provides information on the isotropic and anisotropic splitting of the hole as well as the g-value of the electron. Photoluminescence associated with Si donors in quantum wells was first reported by Shanabrook and Comas [70, 71]. The observation of the Si donor bound exciton in a quantum well was reported by Y. Nomura et al. [72] and Charbonneu et al. [73]. Liu et al. reported on two features interpreted as the exciton bound at the neutral Si donor as well as the exciton bound at the ionized Si donor [74]. Due to the limited binding energy of the confined donor, the energy position of the donor bound exciton is expected to be close to the free-to-bound transition involving the Si donor. This closeness induced some confusion about the interpretation of the observed peaks in luminescence of Si doped quantum wells [70, 71, 74]. Liu et al. [74] made a reinterpretation of the luminescence spectra and performed also a systematic study of the binding energies of the Si donor as well as its associated bound exciton as a function of the well width. Liu et al. also studied the exciton bound at the Si donor as the donor impurities were at different positions within the well and found that the bound exciton binding energy varied in a very limited way, i.e., the exciton binding energy is fairly insensitive to the donor position. This fact can intuitively be understood from the fact that the wave function of the donor electron is extended and more determined by the confinement potential than the donor position. Reynolds et al studied recombination processes, bound excitons and free-to-bound transitions, in wider Si doped quantum wells (200

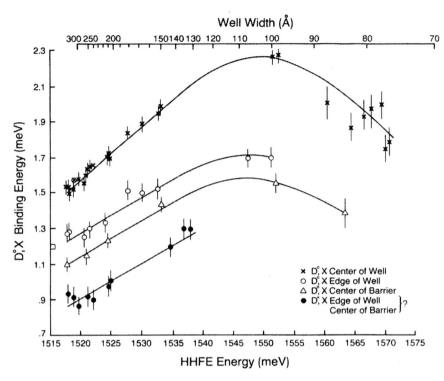

Fig. 4.11. The dependence of the binding energies of the exciton bound to the neutral donors located at the center of the well, the edge of the well and the center of the barrier, respectively, as a function of heavy hole free exciton energy or the well width. (From [77])

and 350 Å) [75–77]. In photoluminescence excitation spectra, they observed a peak at higher energy interpreted as the light hole state of the donor bound exciton. This interpretation is based on energy considerations and magnetic field dependence. They also derived the binding energy of excitons bound to neutral donors in GaAs–AlGaAs quantum wells versus the well widths and the doping positions. The binding energies are found to increase as the well width was reduced until about 100 Å, after which they decreased (see Fig. 4.11).

Well Width Dependence

In photoluminescence spectra, free-to-bound transitions including the neutral donor are observed. Shanabrook and Comas were the first to report the observation of the free-to-bound transitions in luminescence spectra of quantum wells [78]. From such spectra a rough estimate of the donor energy can be performed. As the physical position of the donor is changed in

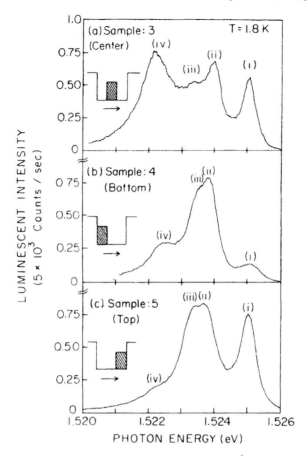

Fig. 4.12. Photoluminescence spectra measured on 210 Å wide quantum wells at 1.8 K with Ar$^+$ laser excitation (488 nm) for a well, which is doped at (**a**) the center of the well, (**b**) the bottom edge and, (**c**) the top edge. The insets represent the nominal doping profile, while the arrows indicate the growth direction. (From [74])

the well, the binding energy of the confined donor is also altered. This can intuitively be understood in a varying overlap between the wavefunction of the bound electron and the donor binding the electron, since the distribution of the electron wavefunction is mainly determined by the confinement potential [21]. Liu et al. [74] observed a shift towards higher energy of the free-to-bound transition involving the donor in luminescence as the donor is moved from the center out to the edge of a 210 Å wide well. This emission blue shift corresponds to a lowering of the donor binding energy, caused by the decreasing wavefunction overlap. Also the intensity of the free-to-bound transition involving the confined donor changes significantly between wells with center donors and wells with edge donors [74] (see Fig. 4.12). Reynolds et al. [75] observed a diamagnetic shift for the free-to-bound transition in the presence of a magnetic field, which was found to be considerably larger than what is observed for excitonic transitions. The observed diamagnetic shift was claimed to have two contributions: one originating from the free hole and the other part originating from the donor. Similarly, a rough estimate of the dependence of the binding energy of the confined donor on the well width can be derived from the energy position of the free-to-bound transitions including the neutral donor in photoluminescence spectra. A quite significant reduction of the donor binding energy (from 13.0 to 10.3 meV) was so-derived as the well width increases from 80 Å to 210 Å [74]. This reduction was found to be in good agreement with theoretical prediction by Greene and Bajaj [30]. Also the excitons bound at the confined donors have been investigated by photoluminescence spectroscopy [74,77,79,80]. The dependence of the bound exciton binding energy on the quantum well thickness has been systematically studied in a number of investigations [48,49,77] (see, e.g., Fig. 4.11).

4.2.4 Selective Luminescence

With access to a tunable laser, the conventional above-bandgap excited luminescence can be replaced by resonant or selective luminescence. Then the laser excitation is tuned into resonance with e.g. the ground state or the excited state of the transition of interest and the luminescence spectrum is monitored as in conventional luminescence. The selective luminescence with resonant excitation can be important in various respects. For instance, the resonant luminescence can be used to sort out the relationship between different emission lines observed in the luminescence spectrum e.g. to identify phonon replicas associated with a certain emission. Another aspect of importance for gaining information on the electronic structure is the resonant excitation with an excited state, which will result in an enhanced luminescence intensity of the ground state emission. Accordingly, this can be seen as a complementary measurement to the photoluminescence excitation spectra. A third essential area is the employment of resonant excitation to enhance the intensity of usually weak satellites in satellite spectroscopy. An example, two particle transitions of an impurity bound exciton with the purpose to get information on the excitation, or so called shake-up processes, will be illustrated below (Sect. 4.2.5).

Satellite spectroscopy is a well established technique for studies of, e.g, the electronic structure of impurities in bulk material. So called two particle transitions of the impurity bound exciton is an example on such a technique, which can reveal minute information on the higher states of the impurity binding the exciton, e.g. a nicely well resolved Rydberg like series excited states of the shallow impurity. In bulk GaAs, satellites of the donor bound exciton are often observed in standard photoluminescence spectra with above bandgap excitation [81]. For other cases, when the satellites are found to be very weak, sometimes below the detection limit, an often employed way to enhance the weak satellites is to perform excitation resonant with the

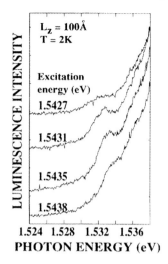

Fig. 4.13. Selective photoluminescence spectra of a 100 Å wide quantum well (the same sample as shown in Fig. 4.9). The excitation energies are in all cases close to the donor bound exciton and are resonant with the bound exciton (at 1.5435 eV) in the second spectrum from the bottom, corresponding to the strongest two-electron satellite

bound exciton [81]. Such satellites, downshifted from the donor bound exciton with an energy separation corresponding to the energy required to excite the donor electron from the donor 1s ground state to excited states. According to applicable selection rules, only transitions to excited states of the same parity as the ground state are allowed. Consequently, this kind of satellite spectroscopy complements the above described infrared methods, in which p-like excited states are monitored. While satellite spectroscopy like the two particle transitions method has been widely used to investigate the electronic structure of donors in bulk material, this technique was not demonstrated for confined donors until 1993 [82]. For moderately Si-doped quantum wells, the exciton bound at the neutral Si-donor will appear in the photoluminescence spectrum in addition to the free exciton with an energy separation of just 1–2 meV. However, no two electron transition satellites are found at above bandgap excitation. Only upon excitation resonant with the donor bound exciton it is possible to observe any two electron transition satellite. Some selective photoluminescence spectra with excitation close to or resonant with the exciton bound at Si donors in the center of a 100 Å wide quantum well are shown in Fig. 4.13. The intensity of the satellite feature observed reaches its maximum, when the excitation is resonant with the donor bound exciton at 1.5435 eV as expected for a two electron transition satellite. The energy separation between the excitation energy and the novel two electron satellite measures to 10.6 meV. This energy separation corresponds to the excitation energy of the donor, since for the case of a two electron satellite, the Si donor

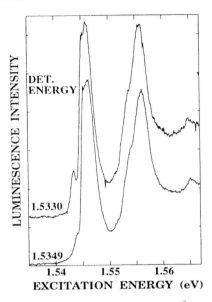

Fig. 4.14. Photoluminescence excitation spectra of a Å wide quantum well (corresponding to the same sample as shown in Fig. 4.9), when the detection energy is resonant and non-resonant with the two-electron satellite shown in Fig. 4.13. For the case of resonant detection, the donor bound exciton appears in this photoluminescence excitation spectrum in addition to the dominating heavy hole and light hole states of the free exciton

is left in its excited 2s state after the recombination instead of the 1s ground state as is the case for the principal bound exciton line. It should also be noted that the satellite is observed also at a laser excitation resonant with either the heavy hole (hh) or the light hole (lh) state of the free exciton as observed in the photoluminescence excitation spectrum, but at a lower intensity level.

A frequently used method to confirm the interpretation of two particle transition satellites is to perform photoluminescence excitation detecting the satellite. For a detected two electron transition satellite, the donor bound exciton is expected to appear in a photoluminescence excitation spectrum. This fact is illustrated in Fig. 4.13 for a 100 Å wide quantum well (with a donor concentration of 5×10^{16} cm^{-3} in the central 20%, corresponding to a sheet density of 1×10^{10} cm^{-2}) with two different detection energies. When the detection energy of this photoluminescence excitation spectrum is resonant with the two electron transition satellite (the upper spectrum in Fig. 4.14), the bound exciton appears in the spectrum in addition to the dominating 1s heavy hole- and light hole-states of the free exciton together with the weaker excited 2s free exciton states related to each of these states at higher energy. The lower spectrum in Fig. 4.14 is a reference spectrum representing a typical photoluminescence excitation spectrum for a non-resonant detection, in which the bound exciton has disappeared and plainly the free exciton

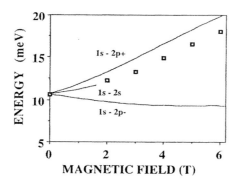

Fig. 4.15. The experimental data on the energy separation between the donor bound exciton and the two-electron satellite observed in selective photoluminescence and photoluminescence excitation measurements in the presence of a magnetic field. Also, the theoretical predictions for the 1s–2p$_\pm$ and 1s–2s donor transition energies are given as solid lines in the figure

states are observed at higher energy. This fact constitutes a confirmation of the interpretation of the observed satellite as being due to a two electron transition satellite process.

In the presence of a magnetic field applied perpendicular to the quantum well layers, the donor bound exciton peak observed in photoluminescence shifts towards higher energy with increasing magnetic field as characterised by a diamagnetic shift rate. While the donor bound exciton line is blue shifted with increasing magnetic field, the two electron transition satellites, on the other hand, are red shifted in the Zeeman measurements, as shown in Fig. 4.15. Accordingly, the donor 1s–2s energy separation, derived from the observed energy displacement of the two electron transition satellites versus the donor bound exciton, increases as a function of the magnetic field, in agreement with theoretical predictions for these donor transitions in the low field regime [83].

4.2.5 Magneto-optics

The Coulombic energy for an effective mass-like impurity as discussed in Chap. 3, can be expressed as

$$E_{\mathrm{b}} = 13.6 \frac{m^*}{\epsilon^2 \, m} \ (\mathrm{eV}) \tag{4.7}$$

while the magnetic energy is

$$\hbar \omega_{\mathrm{c}} = \frac{\hbar \, e \, H}{m^* c} , \tag{4.8}$$

where H is the magnetic field expressed in T, ϵ is the dielectric constant of the material and m^* is the electron effective mass. In the weak field regime, the magnetic field can be treated as a perturbation, while in the strong field regime the Coulombic potential energy is treated as a perturbation. However, for the case of shallow impurities, these two quantities are sometimes comparable, e.g. for donors in bulk GaAs, $\hbar \omega_c \approx 0.87H$ meV and $E_b \approx 5.5$ meV. This fact gives rise to problems in the theoretical approaches, since neither energy scale can be considered to be small relative the other one. Instead, an interpolation procedure has to be introduced for the range between these two extreme field limits. These calculations on the magnetic field dependence for confined donors have been carried out for transitions to p-like excited states in several approaches [24,30], for which experimental magneto-absorption [48,49] and photo thermal ionization spectroscopy (PTIS) [84] results are available. For the corresponding magnetic field dependence of the 1s–2s donor transitions, there are only theoretical predictions reported for the low field domain (< 2 T) [85].

The electronic structure of the donor states is less complex than for the acceptor states. Accordingly, the magnetic field dependence of a hydrogenic donor ground (1s-like) and excited ($2p_{\pm}$-like) states could be theoretically predicted with a reasonable accuracy at an early stage [24,30]. This fact facilitated also the interpretation of the experimental data available. The first experimental progress was consequently achieved for the confined donor states. N.C. Jarosik et al. [48] performed pioneering work on the magnetic field effect on confined donors, making use of far-infrared magnetospectroscopy. According to selection rules for these far-infrared transmission measurements, transitions between states with different parity are allowed, i.e., the transitions between the 1s ground state and the excited $2p_{\pm}$ states are predominanant. N.C. Jarosik et al. [48] were able to observe the blue-shift of the 1s–$2p_{\pm}$ donor transition for a quantum well compared to bulk, but in particular as the well width was reduced as a function of magnetic field applied. The experimentally observed transition energies were compared with the theoretical predictions by R.L. Greene and K.K. Bajaj [30]. The agreement was satisfactory except for the most narrow quantum wells. However, it should be pointed out that the theoretical work was based on the band off-set value of $\Delta E_c / \Delta_g = 0.85$ believed to be the appropriate one at that time, 1985. There were, however, novel results coming up by, e.g., R.C. Miller et al. [86], which indicated that the band off-set value should be modified. This band off-set ratio has some effect also on the donor energies and when instead using a band off-set value of $\Delta E_c / \Delta_g = 0.57$ [86], a better agreement with the experimental results was achieved. Later on, far infrared (FIR) magnetospectroscopic studies have resulted in several reports on e.g. the dependence of the transition energies between the ground state and excited states on the quantum well width and the donor position within the quantum well for the Si donor confined in GaAs/AlGaAs quantum wells [30,86]. An alternative far-infrared technique to study the impurity transitions is the FIR photoconductivity method, which recently has been applied on the confined Si donor [87,88]. In this case, an optically pumped FIR laser is utilized to excite confined donors from the

1s-like ground state to different hydrogen-like excited states in the presence of an applied magnetic field and the photoconductivity is monitored. Due to mixing between states of different parity for donors at an off-center position in the well, parity is not conserved and also 1s-to-ns donor transitions, i.e., $\Delta m = 0$ are observed, but it has been claimed that also $\Delta m = 2$ transitions are observed (in bulk [89]).

4.2.6 Time Resolved Spectroscopy

Part of the incident light in a luminescence measurement is absorbed by the semiconductor. This optical excitation is transferred to the carriers, resulting in a non-equilibrium carrier density, specified by its energy and momentum states. The system will relax towards equilibrium. This means that there is a momentum and energy relaxation of the carriers involved. The momentum relaxation occurs primarily via scattering on a femtosecond time scale, while the energy relaxation of carriers occurs via emission of optical phonons on a picosecond time scale. The subsequent electron-hole recombination finally occurs on a time scale of typically a few hundred picoseconds.

The time resolved measurements related to donors confined in quantum well structures are very limited, due to the small energy separation between the free exciton and the donor bound exciton in GaAs/AlGaAs structures [90–92], resulting a similar decay time. The free exciton and donor bound exciton are thermalized already at 5 K, while the corresponding temperature for acceptors is about 30 K [92]. The radiative free exciton lifetime at 5 K is about 350 ps in wells 50–150 Å wide, while the corresponding radiative lifetime for bound excitons varies between 350 ps and 500 ps in this range. The observed strong temperature dependence of the free exciton and the bound exciton lifetimes is qualitatively explained by detailed balance arguments. Balchin et al. have used the time resolved measurements to identify the transition between the free hole and the electrons bound to Si donors in GaAs/AlGaAs quantum well structures [91]. They found that the lifetime of the Si-donor related free-to-bound is 20 times longer than the lifetime of the heavy hole free exciton. With relative large energy separation between the free exciton and donor bound exciton in II–VI semiconductor structures, the dynamics of exciton relaxation and exciton transfer to donor bound exciton in CdTe/CdMnTe quantum wells have been investigated [93]. It was found that unbound excitons are photoexcited and the timescale of formation of the donor bound excitons varies with excitation intensity from 73 ps to < 8 ps and depends on movement of the unbound exciton in the well and the capture process at the donor site. The actual capture process is characterized by a cross section of 10^{-12} cm^2. When nonresonant excitation densities > 8×10^9 cm^{-2} are used, the donor population saturates and the decay of the unbound and donor bound exciton become thermalized. The decay time of the donor bound exciton were estimated to be about 180 ps for 70 Å wide CdTe/Cd$_{0.85}$Mn$_{0.15}$Te multiple quantum well structures.

5 The Negatively Charged Donor

In analogy with the hydrogen atom binding an extra electron to form the negatively charged H-ion, a donor in a semiconductor might bind another electron to form a negatively charged donor.

5.1 Theoretical Aspects

A negative donor center is formed by a neutral donor binding a second electron to form the negatively charged D^-, i.e. an analogue of the H-ion in atomic physics. The D^- center is of interest in view of interparticle correlations [94–98]. Since there are two electrons bound to the D^- center, singlet and triplet states are formed with the singlet state at the lower energy. These D^- related states can be treated as states of two electrons with an effective mass m^* under a central Coulomb potential screened by the dielectric constant. However, also the electron-electron interaction has to be taken into account for the D^- center. The theoretical predictions have usually resulted in smaller binding energies than the experimental estimates. A better agreement was achieved by the employment of a diffusion quantum Monte Carlo method [94]. The correlation effects become increasingly important in the presence of a magnetic field and/or confinement in a quantum well. A pronounced enhancement of the binding energy of the D^- center is expected, e.g. a sevenfold enhancement for the D^- center in a 100 Å wide quantum well as compared to the value in bulk has been theoretically predicted [94]. In the bulk (3D) case, ionizing transitions from the ground D^- state to free electron Landau levels were first reported by C.J. Armistead et al. [99]. Observations of features related to the D^- center in quantum wells by far infrared magneto-optical measurements have been reported [100, 101]. The observed D^- transition energy was found to increase with increasing electron concentration. P. Hawrylak has performed calculations on the effects of the electron-electron interaction, which resulted in a qualitative agreement with the observed blue-shift [102]. Theoretical calculations of the D^- binding energy were also performed to compare with these experimental results [94–96, 102, 103]. The results show that the binding energy of the D^- center strongly depends on the electron density in the quantum wells. The dissociation energy required to transform the negatively charged D^- to the

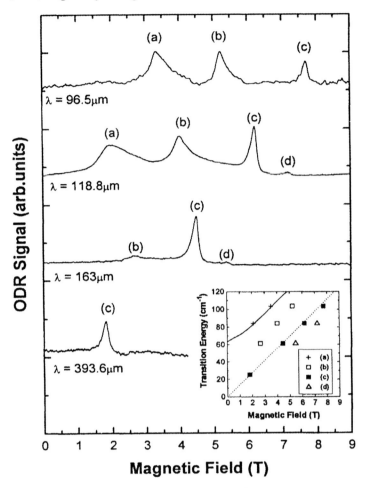

Fig. 5.1. The Optically Detected Resonance (ODR) signal versus magnetic field at four different far infrared (FIR) wavelengths as shown in the figure at 4.2 K with a pumping power of $0.68 \, \text{W/cm}^2$. The observed features are interpreted as (a) the 1s–2p+ transition of the neutral donors (D^0), (b) the singlet transition of the negative donor ions (D^-), (c) the electron cyclotron resonance, and (d) the triplet transition of D^-. The inset shows resulting peak positions vs. the magnetic field, together with a theoretical prediction (*solid line*) for the feature denoted (a) and a *straight dotted line* corresponding to $0.069 \, \text{m}_0$ for the cyclotron resonance. (From [106])

D^0 neutral donor as well as the energy splitting between the D^- states, the D^- triplet and the D^- singlet states, are very small. According to theoretical calculations [94], the binding energy of the D^- center strongly depends on the magnetic field. A similar thermal ionization energy of the D^- center has been concluded from the far-infrared magneto-optical measurements [101].

5.2 Experimental Aspects

The experimental results reported up to now are related to the singlet D^- state, while the corresponding reports on observations of the triplet D^- state are more uncertain. Li et al. [104] claimed to have observed one of the predicted triplet transitions in cyclotron resonance experiments, while Dzyubenko [105] claimed to have searched for triplet transitions in magneto transmission measurements and not found it, due to accidental energy co-incidences with a neutral donor related transition. More recently, Kono et al. [106] reported on far-infrared resonances, including transitions from neutral donors as well as negatively charged donor ions, D^- (see Fig. 5.1). Their results provide evidence for the existence of D^- centers also in illuminated quantum well structures, which are doped solely in the wells, i.e. without any additional modulation doping.

6 Confined Acceptor States

The valence band top of bulk semiconductors with zincblende and diamond structures exhibits a fourfold degeneracy, due to the T_d symmetry, which must be taken into account, when the acceptor states are calculated. Because of the more complex valence band structure, the electronic structure of an acceptor is also much more complicated than for the donor case.

6.1 Theoretical Aspects

While calculations on confined donor states can normally be performed with a single band effective-mass approximation, this one band approximation often fails for the more complicated acceptor states in quantum wells. There are two main reasons for this. Firstly, since the energy separation between the subbands at $k = 0$ is smaller than the acceptor binding energy, the Coulomb coupling between the different subbands must be included in a proper representation of acceptor states. Secondly, while the donor states are mainly associated to one quantum well subband, the full quantum well dispersion has to be taken into account in the confined acceptor case. The acceptor states are associated with the more complex valence band, which is fourfold degenerate including spin at the Γ point for the quantum well case. Accordingly, for a theoretical approach on the confined acceptor, a four band effective mass theory is required. In addition, the non-spherical dispersion of the valence band and the effect of the central-cell corrections have to be considered for the acceptor case.

6.1.1 Effective Mass Approximation

The acceptor Hamiltonian in a quantum well is given by the 4×4 matrix operator

$$H = H^{\text{kin}} + H_{\text{p}}^{\text{QW}} + H^{\text{C}}. \tag{6.1}$$

The H^{C} is the Coulomb potential for an acceptor charge in the center of the well combined with a set of image charges due to the dielectric mismatch of the surrounding layers. H_{p}^{QW} is the quantum well potential. The quantum well potential, H_p^QW, must properly include the effects of the deformation potential and bandgap offset between well and barrier materials. H^kin represents the kinetical energy and is given by the Luttinger–Kohn Hamiltonian [107],

$$H^\text{kin} = \begin{pmatrix} P+Q & L & M & 0 \\ L^+ & P-Q & 0 & M \\ M^+ & 0 & P-Q & -L \\ 0 & M^+ & -L^+ & P+Q \end{pmatrix}, \qquad (6.2)$$

where

$$P = \frac{\gamma_1 \hbar^2}{2m_0} k^2,$$

$$Q = \frac{\gamma_2 \hbar^2}{2m_0} (k_x^2 + k_y^2 - 2k_z^2),$$

$$L = -i\frac{\sqrt{3}\,\gamma_3 \hbar^2}{2m_0} (k_x - i\,k_y)\, k_z,$$

$$M = \frac{\sqrt{3}\,\hbar^2}{4m_0} (\gamma_2 + \gamma_3)\, (k_x - i\,k_y)^2.$$

The γ_1, γ_2 and γ_3 are the Luttinger parameters describing the Γ_8 valence band. The Luttinger Hamiltonian is taken in the axial approximation and the acceptor states can consequently be classified according to the z component of the angular momentum m.

The symmetry of a central acceptor ($z_0 = 0$) in a quantum well is reduced from T_d (in bulk) to D_2d, in which only two representations, Γ_6 and Γ_7, are allowed. For a non-central acceptor ($z_0 \neq 0$), the symmetry group reduces to C_2v with the Γ_5 representation. The quantum well potential lifts the degeneracy of the band also at the Γ point, which gives rise to a splitting between the two Kramers doublets with opposite sign of m but with the same parity under inversion. The doublets can unambiguously be labeled by the m quantum numbers: the $m_\text{J} = \pm 3/2$ (heavy hole) and $m_\text{J} = \pm 1/2$ (light hole) states. However, the complicated mixing between these states at finite $\mathbf{k}$ has to be taken into account. The different theoretical approaches reported are based on the effective mass theory [108–111] or the effective tight binding model [112]. From these calculations, it is deduced that the bulk $1S_{3/2}(\Gamma_8)$ acceptor ground state is split into two doublet states: A more strongly bound state, $1S_{3/2}(\Gamma_6)$, of dominating heavy hole character and another state, $1S_{3/2}(\Gamma_7)$, associated mainly with the light hole subbands, hereafter referred to as the acceptor heavy hole and light hole states, respectively. It should be stressed that an assignment to particular subbands is in principle only applicable for the excited acceptor states, for which the Coulomb coupling between the subbands constitutes only a correction term [110]. For the acceptor ground states, on the other hand, the binding energies are comparable to or even exceed the

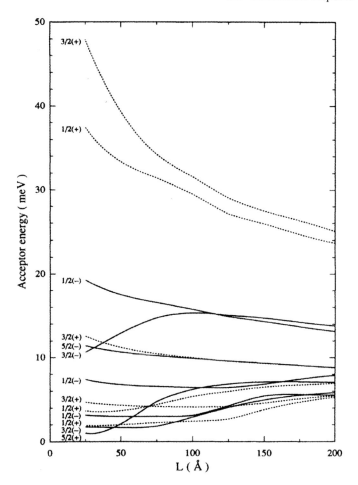

Fig. 6.1. Energy levels for on-center acceptors in $GaAs/Ga_{0.6}Al_{0.4}As$ quantum wells as a function of the well width. The acceptor energies are given as the binding energies with respect to the first heavy-hole subband. The *dashed curves* represent the acceptor energy for states with the symmetry of the two ground states. The *solid lines* correspond to states of different symmetry and are more accurate. The absolute value of the angular momentum m and the parity with respect to inversion indicate the symmetry of the acceptor states. (From [110])

subband separation. These states are in fact strongly mixed. However, the symmetries of the states are very different. Although we will hereafter refer to these states as heavy hole and light hole acceptor states, it should be kept in mind that this assignment to particular subbands is less appropriate for the acceptor states. The binding energies of acceptor Γ_6 and Γ_7 states have been calculated by, e.g., G. Einevoll and Y.C. Chang [112] and S. Fraizzoli and A. Pasquarello et al. [110]. The energies for the total binding energies for the acceptor Γ_6 and Γ_7 ground states as well as the splitting between these states increase with decreasing quantum well width (down to $L_z = 50$ Å), as shown in Fig. 6.1. Also the binding energy for the Γ_6 excited state increases as the quantum well width decreases. The binding energy of the Γ_7 excited state, on the other hand, is predicted to decrease with reduced quantum well width. The reason for this is found in the fact that the Γ_7 excited state is mainly associated with the light hole subbands. If the energy of the Γ_7 excited state is instead given with reference to the first light hole subband, also the energy of the Γ_7 excited state is found to increase with decreasing well width. With the different effective mass calculations, the calculations by S. Fraizzoli and A. Pasquarello et al. [110] show a flexibility in order to taking into account different effects, such as image charges, different dielectric constant and different effective mass in barrier and well layers. The chosen wavefunctions for acceptor states are four-component function,

$$F^m(\rho, \theta, z) = [F^{m,s}] = [F^{m,3/2}, F^{m,1/2}, F^{m,-1/2}, F^{m,-3/2}]. \qquad (6.3)$$

With s tuning over the spin indices $-3/2$, $-1/2$, $1/2$, and $3/2$. m is the hole angular momentum and is a good quantum number. In order to achieve fast convergence for different acceptor states, the s component of an acceptor envelope function of definite angular momentum m can be expanded into a set of basis functions, separable in the coordinates ρ and z:

$$F^{m,s}(\rho, \theta, z) = e^{i(m-s)\theta} \sum A_{nl}^{m,s} \rho^{|m-s|} e^{-\alpha_l \rho} g_n^s(z), \qquad (6.4)$$

where the function g_n^s is chosen to be the s-spin component of the four-component envelope function g_n, which describes a quantum well subband state at $k_\parallel = 0$. It is important to note that the full Hamiltonian of acceptors in center-doped quantum wells has the time-reversal symmetry without applying an external magnetic field. Therefore each acceptor state in quantum well structures is doubly degenerate at zero magnetic field, i.e., states with the same parity (with respect to inversion of the center of the quantum well) and with angular momenta $+ \mid m \mid$ and $- \mid m \mid$ are degenerate. In order to distinguish between the two $\pm \mid m \mid$ states of a doublet, it is useful to consider the reflection operator σ with respect to the $z = 0$ plane. The two states with $\pm \mid m \mid$ of a doublet can be chosen to have opposite parity with respect to σ [110]. By using above given wavefunctions, most of the integrals which appear in the matrix elements of Hamiltonian can be carried out analytically. Particularly, in the calculation of the matrix elements of the Coulomb potential H^C an auxiliary integral which decouples ρ and z-coordinates is introduced using the well-known transformation,

$$\frac{1}{\sqrt{(\rho^2 - z^2)}} = \int e^{-|z|q} J_0(\rho q) \, dq. \qquad (6.5)$$

The calculated results in comparison with the experimental observations will be demonstrated below (Sect. 6.2).

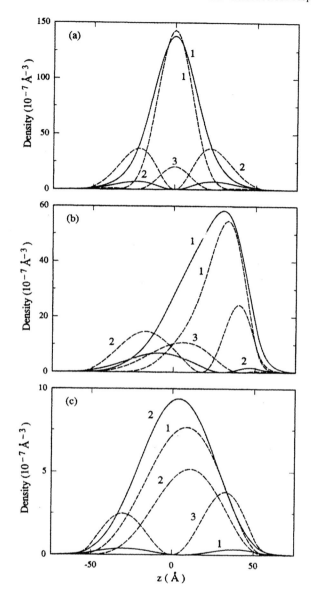

Fig. 6.2. The charge density along the z-axis of the first acceptor states in a 100 Å GaAs/Ga$_{0.6}$Al$_{0.4}$ As quantum well for **(a)** an on-center acceptor ($z_0 = 0$), **(b)** an on-edge acceptor, i.e. at the interface ($z_0 = 50$ Å), and for **(c)** an acceptor at $z_0 = 100$ Å, i.e. in the barrier. The *solid lines* correspond to states with $m = 1/2$, while the *dashed lines* correspond to states with $m = 3/2$. The curves related to states of the same symmetry are numbered following decreasing binding energies. (From [111])

The dipole transition rule is assumed for transitions between acceptor states. The oscillator strength of transitions from the ground state to the excited state is given by

$$f_{i0}^{mn} = 2m_0(E_i^n - E_0^m)/(\hbar^2\gamma_1) \mid \langle F_0^m \mid \boldsymbol{\epsilon} \cdot \boldsymbol{r} \mid F_i^n \rangle \mid^2 , \qquad (6.6)$$

where E_0^m, F_0^m and E_i^n, F_i^n are the energies and envelope functions of ground and excited states, respectively, $\boldsymbol{\epsilon}$ is the polarization vector of the electromagnetic radiation, and γ_1 is the valence band Luttinger parameter.

According to (6.6) and the acceptor wave function given in (6.4), the transition rules for the electromagnetic transitions allow only transitions between acceptor states of opposite parity with respect to inversion. Consequently for the case of $x-$polarization, only transitions between acceptor states with $\Delta m = \pm 1$ and of the same parity with respect to the reflection operator σ are allowed. On the other hand, z-polarized transitions occur only between acceptor states with $\Delta m = 0$ and of opposite parity with respect to the operator σ. The details concerning the infrared absorption spectra of acceptors confined in the center of GaAs/AlGaAs quantum wells have been calculated [110,113].

Also the charge density distribution has been calculated as a function of acceptor position [111]. As the acceptor is moved away from the center of the quantum well, the charge density along the z axis is reduced, since the acceptor states become more extended. The charge density distribution of an acceptor remains confined in the quantum well even if the acceptor is moved away towards the interface (Fig. 6.2). Despite the asymmetric charge distribution for an interface acceptor, the s and p characters of the wave functions remain essentially maintained [111]. If the acceptor is moved further away from the well center into the barrier, the charge distribution of an acceptor remains not only confined in the quantum well, but becomes in fact less asymmetric and is peaking closer to the center of the well again. This effect can be explained by the fact that only the tail of the Coulomb potential is effective, while the confining barriers of the well predominate.

6.1.2 The Presence of an External Field

When an electric field is applied along the growth direction of a quantum well structure, the acceptors confined in the central region of the well layer were theoretically investigated by means of a simple one band model [114]. The binding energy of a shallow acceptor in GaAs/AlGaAs quantum wells was later calculated by means of a variational method. It was found that the electric field affects the difference of the binding energies calculated from a spatially-dependent-screening theory and those from a constant-screening theory. The results show that an electric field has different effects on the acceptor binding energies, when the acceptors are located at different positions in the well layer. The change of the acceptor binding energy is stronger, when the acceptors are away from the center position.

The uniaxial pressure or built-in strain in a lattice mismatched quantum well system strongly modifies the electronic structure of the confined acceptors in quantum well structures. The theoretical and experimental studies of the acceptor ground state in strained quantum well systems were reported by Loehr et al. [109]. They show experimentally and theoretically that the binding energy of the acceptor will decrease with the increase of both the tensional and the compressional strain. Later on, the effective mass calculations of the acceptor energies based on the work by Pasquarello et al. [111] were extended to include stress-effects or built-in strain in lattice-mismatched quantum well system [115].

In order to include the strain effects in a proper way, the quantum well potential in (6.1), H_p^{QW} must include the effects of the deformation potential. That is, H_p^{QW} will contain a square-well potential, $H_{hh,lh}^{QW}$, for the heavy hole (hh) and the light hole (lh), and includes a potential difference (V_p) between the heavy hole and light hole band edges in the well (due to the built-in or the external strain) [115]:

$$H_p^{QW} = H_{hh,lh}^{QW} + V_p . \qquad (6.7)$$

Note that $H_{hh,lh}^{QW}$ may have different values for the heavy hole and light hole states, when the deformation potential is included. In the case of strain-free quantum wells (i.e. the barrier materials and the well material have a similar stress coefficient), an applied external stress only changes the energy separation (V_p) between the heavy hole and light hole state and $H_{hh,lh}^{QW}$ will not change. The splitting between the heavy hole and light hole states due to an applied pressure can be described by $V_p(\pm 3/2) = -V_p(\pm 1/2) = D$ in (6.7). In this definition, a negative D means that the stress potential has a compressive character. In that case, the heavy hole state will be the ground state. On the other hand, when D is positive, we have a tensional strain situation, and the light hole state will be the ground state. If the barrier and well layers have different stress coefficients, the stress dependence of $H_{hh,lh}^{QW}$ must also be treated in a proper way. In the case of lattice mismatched quantum well systems, it is clear that the built-in strain influences both the heavy hole and light hole band offset $H_{hh,lh}^{QW}$ and the energy separation V_p in (6.7).

From these calculations, the detailed energy levels of acceptors and oscillator strengths of the transitions between the acceptor ground and different excited states in quantum well structures in the presence of stress perturbations were obtained and can be compared with experimental data. The effective mass model, containing all major effects such as dielectric difference between barrier and well layers and continuous states, is employed in order to obtain reliable energies of the acceptor excited states. The results show that the deformation potential has a stronger influence than the confinement potential, on the acceptor ground state splitting relative to the first heavy hole–light hole subband splitting. Figure 6.3 shows the even symmetric acceptor states, in which the solid and dashed curves correspond to states with $(3/2, +)$ and

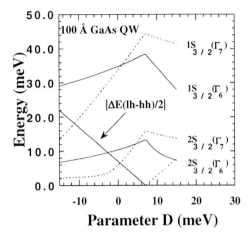

Fig. 6.3. Pressure dependence of the binding energies of the even parity acceptor states for an on-center impurity in a 100 Å wide GaAs/Ga$_{0.7}$Al$_{0.3}$As quantum well. The *solid* and *dashed* curves correspond to states of the $(3/2, +)$ and $(1/2, +)$ symmetry, respectively. The binding energies are given with respect to the bottom of the first hole subband. The energy separation $|\Delta E(\text{lh} - \text{hh})|$ between the first light hole and the first heavy hole sublevel is also indicated in the figure. The solid line for $|\Delta E(\text{lh} - \text{hh})|$ corresponding to the ground state of heavy hole character and the dashed line is the corresponding line for the ground state of light hole character

$(1/2, +)$ symmetry, respectively. The fourfold degenerate acceptor ground state in bulk, $1S_{3/2}(\Gamma_8)$, splits into two twofold degenerate states, $1S_{3/2}(\Gamma_6)$ and $1S_{3/2}(\Gamma_7)$, for acceptors located at the center of the quantum well. The $1S_{3/2}(\Gamma_6)$ and $1S_{3/2}(\Gamma_7)$ states are related to the heavy hole ground state of $(3/2, +)$ symmetry and the light hole ground state of $(1/2, +)$ symmetry, respectively. The acceptor energies are given with respect to the bottom of the first hole subband, which corresponds to the energy of the lowest hole level in an impurity-free quantum well. Worthy to note is the fact that (depending on the strength and the sign of the deformation potential and the quantum well confinement) the ground state can be either the heavy hole or light hole state. In the case of a 100 Å GaAs/Al$_{0.3}$Ga$_{0.7}$As quantum well, the cross-over between the first heavy hole and the first light hole sublevels occurs at a value of $D = 7.0\,\text{meV}$ (Fig. 6.3). It is also of interest to note that the cross-over between the ground acceptor states, $1S_{3/2}(\Gamma_6)$ and $1S_{3/2}(\Gamma_7)$ occurs before the cross-over of the first heavy hole and light hole subbands (Fig. 6.3). For the case of acceptors confined in InGaAs/AlGaAs and GeSi/Si quantum well structures, the effects of the built-in strain on the electronic structure of the acceptors and the oscillator strenghts of the transitions between acceptor sub-levlels have been calculated [115]. The presence of an external magnetic field along the growth direction can further break the symmetry of the acceptor

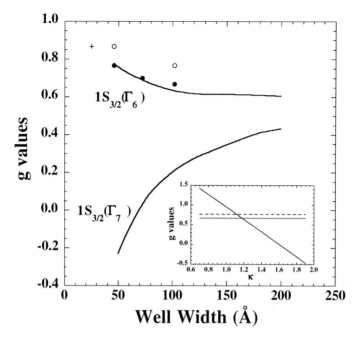

Fig. 6.4. The g values of the 1s acceptor states calculated for varying well widths. The inset shows the g-value (*solid line*) of the acceptor $1S_{3/2}(\Gamma_6)$ states for a 100 Å wide quantum well calculated as a function of the Luttinger parameter, κ. The *unfilled circles* are the experimental data from D.N. Mirlin et al. [126], the *solid circles* from V.F. Sapega et al. [124], and the *crosses* refer to data from H.W. van Kesteren et al. [125]

states confined in the quantum well structures. The magnetic field effects are included, if we replace $k = -i\nabla + |e| A/(\hbar c)$ in (6.2) with the Hamiltonian from (6.1) including the term $2\mu_B\kappa\mathbf{B} \bullet \mathbf{J} + q\mu_B(B_x J_x^3 + B_y J_y^3 + B_z J_z^3)$, where $\mathbf{B} = \nabla \times \mathbf{A}$, and $\mathbf{J}$ is the angular momentum of the hole [113,116].

The fourfold degenerate acceptor ground state in bulk, $1S_{3/2}(\Gamma_8)$, is split into two twofold degenerate states, $1S_{3/2}(\Gamma_6)$ and $1S_{3/2}(\Gamma_7)$, when the acceptors are located in the center of the quantum well, at zero magnetic field. The $1S_{3/2}(\Gamma_6)$ and $1S_{3/2}(\Gamma_7)$ states are related to the heavy-hole ground state of $(3/2, +)$ symmetry and the light-hole ground state of $(1/2, +)$ symmetry, respectively. When a magnetic field is applied along the z direction, the degeneracy is further lifted. This Zeeman splitting is larger for the heavy-hole related $1S_{3/2}(\Gamma_6)$ ground states than for the light-hole related $1S_{3/2}(\Gamma_7)$ ground states. The corresponding Zeeman splitting for the excited 2S heavy hole states is much smaller than for the 1S heavy hole ground states, while the opposite is true for the light hole acceptor state. Similarly, the p-like excited states also split in the presence of an applied magnetic field. The splitting of both s-like and p-like symmetric acceptor levels are generally not linearly dependent on the magnetic field, especially for the excited states [113, 116].

From the calculations, the g-factors of each acceptor state can be deduced. The g-factor is a parameter to describe the linear splitting of acceptor states with an applied magnetic field. The splitting of the $\pm \mid J_z \mid$ states can be written as

$$\Delta E_m = \mu_B \, g_{J_z} B[\mid J_z \mid - (- \mid J_z \mid)] = 2\mu_B \, g_{J_z} \mid J_z \mid B \qquad (6.8)$$

or alternatively

$$\Delta E_m = \mu_B \, g_{J_z} B , \qquad (6.9)$$

where J_z is the magnetic angular momentum along the direction of the magnetic field. g_{J_z} is a g-value related to the angular momentum J_z. Obviously, the definitions used in (6.8) and (6.9) differ by a factor $2\mid J_z \mid$, which means that for the $\mid \pm 3/2 \rangle$ state, the g-values given by (6.8) and (6.9) differ by a factor of 3, while for the $\mid \pm 1/2 \rangle$ states, the two definitions are identical. It is important to note that the system discussed is the symmetric center doped quantum well, i.e. the $\mid \pm m \rangle$ states are degenerate without an external magnetic field. If the symmetry is low enough to lift such a degeneracy at zero magnetic field, the definition (6.8) is not valid anymore. From the definition given in the above equations together with the magnetic field dependence of the calculated acceptor energy levels, the g-values have been deduced (see Fig. 6.4) [117].

6.2 Experimental Aspects

Information concerning the electronic structure of acceptors can be experimentally obtained from optical spectroscopy such as infrared absorption, photoluminescence and Raman scattering measurements. The infrared absorption measurement, e.g. by Fourier Transform Infrared (FTIR) spectroscopy, is a direct way to measure the energy separations between the acceptor ground and the different p-like excited states. The rich information concerning the electronic structures of acceptors from infrared absorption measurements has been nicely demonstrated in bulk GaAs [118], where the far-infrared photoconductivity measurements were used. From these measurements, the transitions $1S_{3/2}$-$2P_{3/2}$, $1S_{3/2}$-$2P_{5/2}(\Gamma_8)$ and $1S_{3/2}$-$2P_{5/2}(\Gamma_7)$ were observed and their corresponding transition energies have been experimentally deduced. Consequently, detailed information on the electronic structure of acceptors can be achieved in this way. However, the situation for quantum well structures is more difficult [49]. Due to a small absorption volume of the acceptor doped quantum well structure, the infrared measurements are difficult to perform and the so far derived experimental information is still limited [49].

The most experimental works on acceptors confined in quantum well structures were carried out by means of photoluminescence, selective photoluminescence and Raman scattering measurements. Based on our knowledge

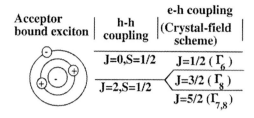

Acceptor bound exciton	h-h coupling	e-h coupling (Crystal-field scheme)

$$J=0, S=1/2 \qquad J=1/2 \ (\Gamma_6)$$

$$J=2, S=1/2 \left\langle \begin{array}{l} J=3/2 \ (\Gamma_8) \\ J=5/2 \ (\Gamma_{7,8}) \end{array} \right.$$

Acceptor

$$J=3/2 \qquad 1S_{3/2}(\Gamma_8)$$

Fig. 6.5. Schematic picture of the states for an acceptor and its bound exciton at different perturbation conditions in bulk GaAs

from optical studies of bulk semiconductors, a neutral acceptor can bind an electron-hole pair to form a so called acceptor bound exciton (see Fig. 6.5). The impurity-related emissions e.g. the acceptor bound exciton and the transition between free electrons and holes bound to acceptors, so called free to bound transitions, are referred to as extrinsic luminescence to distinguish from emissions of intrinsic origin such as the free exciton. A pioneer work on the extrinsic luminescence related to acceptors in GaAs/AlGaAs was reported by R.C. Miller et al. [119].

6.2.1 Infrared Measurements

The most pertinent work in the infrared range has been accomplished by Fourier Transform Infrared spectroscopy. Similarly to the donor states described above (Sect. 4.2.1), the transitions between the s-like ground state and the excited p-like states are predominant in such infrared spectra. A.A. Reeder et al. [49,120] performed transmission measurements on Be acceptors confined in quantum wells with various thickness. The transitions between the $1S_{3/2}$ ground state and the excited $2P_{3/2}$ states dominate the far infrared spectrum, but also transitions to the excited $2P_{5/2}$ states are clearly resolved. Due to the different selection rules prevailing in the infrared transmission experiments compared to, e.g., resonant Raman experiments or two-hole transitions of the acceptor bound exciton, the results from infrared transmission constitute an important complement to these measurements. Furthermore, the results are essential for comparison with theoretical predictions.

6.2.2 Raman Scattering

The pioneer work on electronic Raman scattering performed on impurities confined in quantum wells was published 1984 by B.V. Shanabrook et al. [121]. They investigated the transition between an electron bound at a confined donor and a hole from the valence band. For excitation above the bandgap, the normal free-to-bound transition in photoluminescence is observed, but if the excitation is down-shifted in energy, below the bandgap, towards the excited states of the donor, the observed structure becomes more narrow and is also red-shifted. The free-to-bound photoluminescence is transformed into an electronic Raman scattering. The optimum conditions for this transformation with an associated intensity maximum occurs, when the excitation is resonant with the 2S state, consistent with the selection rules applicable to Raman scattering.

Later on, the same group monitored also resonant donor states originating from higher conduction subbands in resonant Raman scattering experiments performed on Si center doped quantum wells [122]. The Raman scattering measurements were performed at an energy corresponding to the optical gap $E_0 + \Delta_0$ of GaAs, since the scattering involving states of the conduction band is enhanced at this energy [123]. Upon excitation resonant with the optical gap $E_0 + \Delta_0$ of GaAs, a characteristic doublet structure was observed. This doublet was interpreted as due to transitions from the ground state of the Si donor in the lowest subband to resonant donor states in higher subbands.

6.2.3 Hole g-Values

The computations on magnetic field dependence on the acceptor states are relatively complicated. Due to this complexity, the number of papers dealing with the effect of a magnetic field is limited. An early theoretical treatment of the Zeeman effect on an acceptor state was reported by Masselink et al. [108]. They predicted the ground state splitting between the acceptor Kramers' doublet states, the $m_j = \pm 3/2$ states, however, the effect of confinement was however not taken into account (the effective L_z and J_z were seen as practically identical to that of the bulk acceptor and the effect of the confinement was accordingly negligible). Consequently, the same splitting as for a bulk acceptor was predicted, which was significantly larger than the experimental results reported [124–126].

S. Frazzioli and A. Pasquarello performed calculations on the acceptor transitions in the absence of a magnetic field perturbation [110]. In these calculations, the continuum states are included in the basis set, in order to improve the accuracy for the excited states of the acceptor. The theory by Frazzioli and Pasquarello was later extended to include a magnetic field perturbation [117]. This approach has the advantage compared to former computations that also the acceptor transitions involving excited states exhibit a good agreement with the experimental results. The g-factors for the shallow acceptor were calculated for the $1S_{3/2}$ ground state and the $2P_{3/2}$ excited state for a comparison with the experimentally observed transitions involving these states [117].

On the experimental side, the g-tensor of the acceptor bound hole has been determined by spin-flip Raman scattering. This method has earlier been employed on bulk materials [127, 128] and important information on exchange interactions and g-factors of impurities and free carriers has been achieved. The formation of an exciton can occur via the allowed absorption process of either σ^- polarized light (forming a $|-1/2, +3/2\rangle$ exciton state or σ^+ polarized light (forming a $|+1/2, -3/2\rangle$ exciton state. The corresponding formation of $|+1/2, +3/2\rangle$ or $|-1/2, -3/2\rangle$ exciton states is forbidden. Anyhow, these transitions can experimentally be observed. Such transitions require, however, the transformation of the hole state from the $|+3/2\rangle$ to the $|-3/2\rangle$ state, i.e. a flip of the hole angular momentum. In a typical spin-flip Raman scattering experiment, the laser excitation is kept resonant with the bound exciton and the experiment is performed in the presence of a variable magnetic field. Since the transformation can go in both directions, $|+3/2\rangle \rightarrow |-3/2\rangle$ or $|-3/2\rangle \rightarrow |+3/2\rangle$, the spin-flip lines are observed as either a Stokes component on the high energy side or an anti-Stokes component on the low energy side of the laser excitation. These components are strongly circularly polarized, but also temperature dependent [124]. In quantum well structures, the efficiency of the spin-flip Raman scattering is expected to be enhanced due to an increased exchange interaction [129, 130] and oscillator strength of the exciton [131] and caused by the confinement. Sapega et al. have observed corresponding strong spin-flip Raman scattering in acceptor doped quantum wells as well as in undoped structures (Fig. 6.6). The observed Stokes and anti-Stokes are strongly polarized and the Raman shift, $\hbar\omega$, is linearly dependent on the magnetic field, B, applied according to

$$\hbar\omega = g\,\mu_0\,B\cos(\phi)\,, \tag{6.10}$$

where μ_0 is the Bohr magneton, g is the g factor of the acceptor bound hole and ϕ is the angle between B and the z direction of the quantum well. For the case of acceptor doped structures, the observed spin-flip Raman scattering originates from a spin-flip of the hole bound at the acceptor via exchange interaction with a neighbouring photoexcited exciton in the presence of a magnetic field. The transition, which gives rise to the spin-flip Raman scattering process, is between states with different angular momenta, $|+3/2\rangle \rightarrow |-3/2\rangle$ or $|-3/2\rangle \rightarrow |+3/2\rangle$, of the magnetically split acceptor ground state. The corresponding spin-flip scattering in undoped quantum wells is due to the angular momentum flip of a localized exciton via interaction of acoustical phonons. From the dependence of the Stokes and anti-Stokes replicas due to the spin-flip Raman scattering, the longitudinal component of the g factor of the acceptor bound hole has been determined ($g_\parallel = +2.3$ for $L_z = 40\,\text{Å}$ to be slightly decreasing with increasing well width [124]). The experimental re-

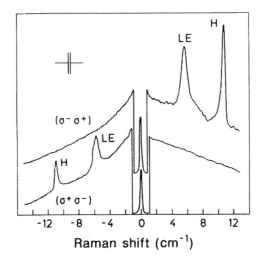

Fig. 6.6. Raman spectra of acceptors confined in a GaAs/AlGaAs quantum well, measured in the (σ^-, σ^+) and (σ^+, σ^-) configurations in a magnetic field of $B = 10\,\mathrm{T}$ using an excitation energy of $1.628\,\mathrm{eV}$. The H label corresponds to the hole spin-flip Raman line, while LE corresponds to the localized exciton angular momentum flip line. (From [124])

sults have been compared with the theoretical predictions [117], with a good agreement obtained (see Fig. 6.4).

6.2.4 Luminescence Measurements

The radiative recombination as can be detected in luminescence measurements can provide essential information about the properties of the host material as well as about the impurities in the material. Depending on the excitation process, we talk about e.g. photo-, cathodo-, electro- and thermoluminescence. In this book, the first one, the photoluminescence, is the predominant luminescence technique referred to. Among many advantages, the possibility to tune the excitation intensity and energy and the excess to ultrashort laser pulses for time resolved luminescence measurements, should be mentioned. By photoluminescence spectroscopy, one can gain extensive information on essential properties of the quantum well. The structure is normally photo-excited by a laser and the resulting luminescence emission is monitored and spectrally analyzed. To gain further information, parameters like temperature, excitation intensity and energy can be varied. In addition, luminescence spectroscopy perturbed by e.g. an electric field, a magnetic field, a microwave field can further elucidate the characteristics of the quantum well.

6.2.5 Selective Photoluminescence and Excitation Spectroscopy

The possibility to choose a specific excitation energy, opens fascinating possibilities within luminescence spectroscopy. For instance, weak satellites can be enhanced, associated emission lines can be connected, the identification of an excited state can be verified, by comparing a selective photoluminescence spectrum using resonant excitation with a reference spectrum with non-resonant excitation, as will be exemplified below. But also luminescence spectra with excitation above the barrier energy can be significantly different from a corresponding spectrum with excitation within the well. From such comparisons, valuable information on e.g. capture processes and competing recombination processes can be obtained. The excitation spectroscopy can provide useful information on the excited states associated with a particular recombination transition. The photoluminescence excitation spectrum is obtained by scanning the excitation photon energy, while monitoring the intensity at the selected emission energy.

6.2.6 Recombination Processes

The fundamental recombination process in an intrinsic semiconductor is the band-to-band free carrier transition, i.e., a free electron from the conduction band recombining with a free hole from the valence band. However, due to the mutual Coulomb interaction between carriers of opposite charge, an electron-hole pair, an exciton, is formed with a hydrogen-like resonance below the bandgap. The free exciton binding energy can be estimated by making use of the effective-mass approximation:

$$E_b^{exc} = 13.6 \frac{\mu}{m_0 \epsilon^2} \text{ (eV)},$$

where μ is the reduced mass of the exciton:

$$\mu = \frac{m_e m_h}{m_e + m_h} m_0 .$$

The binding energy of the free exciton in a quantum well is significantly larger (typically a factor 2–3) than the corresponding free exciton in the 3D case. The free exciton will accordingly survive to higher temperatures in quantum wells and is found to dominate the luminescence spectrum of undoped quantum wells, not only at low temperatures, but also at elevated temperatures the free exciton recombination is strong relatively the band-to-band transition. For instance, J.P. Bergman et al. [134] found that for a 100 Å wide GaAs/AlGaAs quantum well, the free exciton emission will be predominant even at room temperature (about 30% higher luminescence intensity than the free carrier recombination).

The first reports on luminescence from GaAs/AlGaAs quantum well structures referred to undoped material and the resultant luminescence was accordingly mainly of intrinsic origin [134–137]. As impurities are introduced in

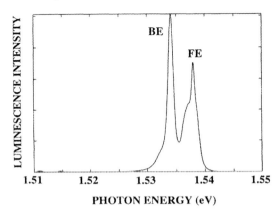

Fig. 6.7. Photoluminescence spectrum of an acceptor doped $(3 \times 10^{16}\,\text{cm}^{-3})$ 150 Å wide quantum well, measured at 1.6 K with an excitation energy of 1.652 eV. The free exciton (FE) and acceptor bound exciton (BE) dominate this spectrum, but the biexciton is also observed 1.1 meV below the free exciton. The intensity of the biexciton versus the free exciton increases with increasing excitation intensity

the lattice there is an increasing probability that the free exciton is trapped by an acceptor/donor to form a bound exciton.

Photoluminescence spectra of in particular wide quantum wells often exhibit overlapping spectra of the quantum well associated emissions and the bulk spectrum originating from the buffer layer and/or substrate. This fact constitutes a complicating factor for the spectroscopic analysis. By utilizing selective photoluminescence, the origin of the overlapping spectra can be distinguished. Similarly, selective photoluminescence can be used to spectroscopically separate the emissions from quantum wells of different width.

The PL spectra with above bandgap excitation for acceptor doped quantum wells measured at low temperatures are completely dominated by the free exciton and the acceptor bound exciton as illustrated in Fig. 6.7 for a center doped 150 Å wide quantum well. In addition, a novel peak shows up between the free and bound excitons, 1.1 meV below the free exciton. This novel feature will gain intensity considerably with increasing laser excitation intensity and is interpreted as the biexciton [132, 133].

The probability for a free exciton to become trapped at an impurity is smaller in a quantum well compared to the bulk case. This is partly due to the shorter life time of the free exciton in a quantum well and consequently a reduced probability to get trapped before recombination. Accordingly, higher doping levels are required for a predominant bound exciton in comparison with the bulk case. While the predominant luminescence is of extrinsic origin also in undoped bulk material, the situation is strikingly different in quantum well structures with a dominating intrinsic luminescence not only in undoped quantum wells, but also prevailing in moderately doped material. This effect

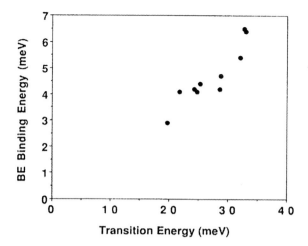

Fig. 6.8. The dependence of the Be acceptor bound exciton binding energy on the transition energy for the $1S_{3/2}(\Gamma_6)$-$2S_{3/2}(\Gamma_6)$ transition for the acceptor binding the exciton. The transition energies are deduced from two-hole transition experiments. The point at lowest energy (2.9 meV) corresponds to bulk GaAs. As can be seen in this figure, there seems to be an almost linear relationship between the acceptor transition energy and the bound exciton binding energy. This implies that a correspondence to Haynes's rule in the bulk should also be valid for this quantum well system

is particularly pronounced in narrow quantum wells, while the relative luminescence of extrinsic origin gains successively intensity as the well width is increasing.

The binding energy for the confined acceptor bound exciton, defined as the energy separation relatively the free exciton, is slightly larger than for the corresponding acceptor bound exciton in bulk ($E_b \approx 3$ meV). The dependence of the binding energy for the confined acceptor bound exciton on the degree of confinement, i.e. on the well thickness, has been studied by Holtz et al. [138]. As expected, an increasing binding energy was found as a function of decreasing well width corresponding to an increasing acceptor 1s–2s transition energy, as demonstrated in Fig. 6.8. However, when stating these dependencies, one has to be cautious, since these conclusions are all based on the presumption that the acceptor is located at the center of the quantum well. The dependence of the exciton energy on the acceptor site is significantly more noticeable than the dependence on the well width. For instance, it is found that the bound exciton energy is reduced from 4.2 meV to 1.6 meV, as the acceptor is displaced from the center out to the interface in a 100 Å wide quantum well [139]. This fact can be intuitively understood from the reduced overlap of the wave functions of the exciton and the acceptor binding the exciton, respectively, as the acceptor is moved away from the center.

A pioneer work on the extrinsic luminescence in GaAs/AlGaAs quantum wells was reported by R.C. Miller et al. [140]. They found that the extrinsic and intrinsic luminescence intensities were comparable up to acceptor concentrations of $\approx 10^{17}\,cm^{-3}$. The predominant recombination of extrinsic origin was interpreted as due to a free-to-bound process between a $n = 1$ electron from the conduction band and an acceptor bound hole. The energy position of this free-to-bound band for various quantum wells enabled R.C. Miller et al. [140] to estimate the acceptor binding energy as a function of the well width, L_z. For a narrow quantum well, an additional shoulder was interpreted as a free-to-bound transition involving acceptors close to the interface. Such a free-to-bound transition occurs at a higher energy, since the binding energies for interface acceptors are smaller relatively acceptors in the center of the quantum well according to theoretical predictions, the first one reported by G. Bastard [21].

The intensity of the free-to-bound recombination has been found to depend on several parameters. For instance, Muraki et al. [141] observed a striking oscillatory behavior as a function of the well width with the excitation energy kept above the barrier bandgap. The intensity of the free-to-bound recombination varied by as much as two orders of magnitude between the optimum well width conditions and the "anti-resonance" conditions. This striking intensity was attributed to the electron capture efficiency from the barrier into the well, which exhibit a similar oscillatory behavior, when the well width is varied. Brum and Bastard have calculated the capture time of the electrons via emission of LO phonons as a function of the well width [142] and found as the result an oscillatory behavior as a function of the well width but with two series of resonance's of different origin: One resonance associated with a virtual state, occurring when the highest bound state in the well is resonant with the conduction band edge of the barrier and a second kind of resonance involving LO phonon interaction, occurring when the energy difference between the conduction band edge and the highest bound state matches with the LO phonon energy. The well width dependence of the electron capture efficiency was experimentally verified by means of cw-photoluminescence [143] and later on by time resolved luminescence in the fs range [144].

When the laser excitation energy is tuned down below the barrier bandgap, the intensity of the free-to-bound recombination was found to be strikingly enhanced upon excitation resonant with any of the free exciton states, FE[hh] or FE[lh] [145]. In this case, the enhancement was explained in terms of an Auger-like process, the excitonic Auger. The hole of the free exciton is attracted and captured by an ionized acceptor. The excess energy from this capture process is transferred to the electron, which becomes excited into the continuum. This electron can be monitored directly via photoconductivity measurements [146–148] or by an enhanced intensity of the subsequent free-to-bound recombination between this electron and the hole captured by the acceptor.

An interesting experiment to make use of the delocalized wave function in k-space of an acceptor has been demonstrated by J.A. Kash et al. [149, 150]. In a hot photoluminescence experiment, the recombination between hot electrons (at a certain in-plane wavevector, k) from the conduction band and holes originating from an acceptor is monitored, while the laser photon excitation is tuned to resonance with a hot hole with essentially the same in-plane wavevector, k. By measuring the photoluminescence energy, $\hbar\omega_{PL}$, as a function of the laser excitation energy, $\hbar\omega_{exc}$, the dispersion of the valence band can be evaluated from the differential energy, $\hbar\omega_{exc} - \hbar\omega_{PL}$. It should be noted that this hot luminescence is extremely weak, typically six orders of magnitude weaker intensity than the predominant band edge luminescence in a quantum well. This strong band edge luminescence also hampers the determination of the valence band dispersion at small wavevectors. The so derived results, which can be provided with a meV accuracy, constitute important input information on the subband dispersion for calculations of the complex valence band structure.

6.2.7 Two-Hole Transitions of Bound Exciton

When a photon generated electron-hole pair is trapped by a defect, it is denoted a bound exciton. If the defect is a donor or an acceptor, we normally refer to the donor and the acceptor bound exciton, respectively. For the case of the neutral acceptor, we encounter a system involving two $j = 3/2$ holes and one spin $1/2$ electron. According to general j–j coupling, the two $j = 3/2$ holes couple to form states of a total angular momentum of $J = 0$ and $J = 2$ with the $J = 2$ state at lowest energy [151]. Each of these states is then coupled to the $s = 1/2$ electron to give three states with a total angular momentum of $J = 1/2, 3/2$ and $5/2$, respectively [152]. The electronic structure of the acceptor bound exciton in a quantum well structure is illustrated in Fig. 6.9.

When the acceptor BE recombines, the acceptor is usually left in its ground state, but there is a small probability that the acceptor is instead left in an excited state. If the final state of the acceptor is the ground state, the transition is denoted as the principle bound exciton recombination. If the final state is the excited state of the acceptor, the process is referred to as the two-hole transition or the "two-particle" transitions [153]. The resulting two-hole transition can be monitored as a satellite in a photoluminescence spectrum. Usually the intensity of this satellite is weak, but can be enhanced by resonant excitation. The observation of these satellite transitions are of great importance for the identification of the host impurity at which the excitons bind. In the two-hole transition, the hole of the acceptor is left in an excited state after the recombination instead of in the ground state as is the case for the principal bound exciton recombination. This means that the satellite due to the two-hole transition is red shifted relatively the bound exciton with an energy corresponding to the energy needed to excite the

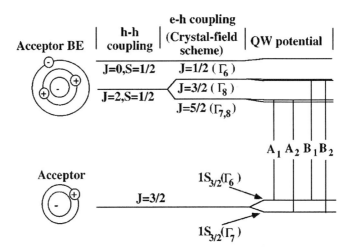

Fig. 6.9. A schematic picture of a confined acceptor and its bound exciton at different perturbation conditions. The possible transitions related to this acceptor bound exciton confined in a quantum well are also indicated in the figure

acceptor hole from the ground state to an excited state. This two-particle transition is best described in terms of a direct process, where its magnitude is determined by the fraction of the final state wave function in the initial state of the same particles in the bound exciton [154].

The method of two hole transitions has been widely employed for acceptor bound excitons in bulk material to gain information on the electronic structure of the exciton, but more importantly the acceptor binding the exciton [152]. The basic idea is to determine the energy separation between the principal bound exciton and the satellite: When the bound exciton recombines, the luminescence spectrum is dominated by the principal bound exciton emission. However, there is a small, but non-zero, probability that acceptor becomes excited to a higher state upon the excitation. This experimentally observed satellite in the luminescence spectrum is usually a weak satellite, red-shifted relatively the principal bound exciton with an energy separation corresponding to the energy required to excite the acceptor from its ground state into an excited state. Satellite spectra were first reported by P.J. Dean et al. for donors in GaP [153], for which the interpretation was confirmed by a comparison with the donor binding energies as derived from donor-acceptor pair spectra. The most fascinating two particle satellite spectroscopy has been demonstrated for the II–VI compounds e.g. hole transitions to s-states with an orbital quantum number up to 10 has been monitored for acceptor bound excitons in ZnTe [155].

The selection rules applicable to two-particle transitions allow transitions connected to excited states with the same parity as the ground state, i.e. s-like excited states, as has been clearly shown for donors in bulk Si [156].

Later on, very accurate determination of the binding energies of the s-like excited states relatively the ground state have been demonstrated for both donors and acceptors also in GaAs.

In practise, the low probability for the process of two hole transitions results in a satellite below the detection limit for the case of above bandgap excitation. A trick commonly used in bulk material to enhance the two hole transitions satellites, is an excitation resonant with the acceptor bound exciton, since the satellites can be regarded as a generalized form of bound exciton recombination. The same trick is applicable also to quantum wells, in which an enhancement of the two hole transitions satellites can also be achieved by excitation resonant with the free exciton [138]. This is illustrated in Fig. 6.10 for tha case of a 69 Å wide quantum well, δ-doped with acceptors in the center of the well. The wells are δ-doped in the middle of the well in order to avoid broadening of the satellite lines caused by the dispersion of the acceptor binding energies. With the laser excitation close to the acceptor bound exciton, additional resonant phenomena can be observed. In addition to the usual two hole transitions satellite, there are novel well defined features "following" the laser excitation, in such a way that the energy separation between the laser excitation and the satellites remains almost constant, as shown in Fig. 6.10. This additional set of satellites is interpreted as originating from a Resonant Raman Scattering process [138].

The same selection rules apply for two hole transitions and Resonant Raman Scattering: Excitation processes to states with the same parity as the ground state are allowed, i.e., transitions to S-like excited states will dominate the spectrum. This is clearly demonstrated in Fig. 6.11, measured for the case of a wider quantum well ($140\,\text{Å}$) in which the satellite corresponding to the 1S–2S acceptor transition state by far dominates the satellite spectrum, but a weak feature due to a transition to the next S-like state, the 3S state, appears at $1.511\,\text{eV}$. In this satellite spectrum, not only the predominant excitations to S-like excited states, but also an additional feature interpreted as the forbidden transition to a P-like excited state, the $2P_{3/2}$ state. It should be noted that for the case of acceptors in bulk GaAs, no two hole transitions satellite originating from a transition to the $P_{3/2}$ state is reported. In addition, there is no proper theoretical approach on the selection rules applicable for two hole transitions up to now. This means that no guidance is derived from the theory, so the prevailing knowledge on these transitions is rather deduced from experimental considerations. When going from bulk to a quantum well, the symmetry is reduced: From T_d for an acceptor in bulk to D_{2d}, for an acceptor in the center of a quantum well, as desribed above in Sect. 6.2. Accordingly, the selection rules are expected to be relaxed for the case of quantum well transitions. The interpretation is supported by comparing the achieved results with findings from other techniques, in which othe selection rules are applicable. For instance, the transition energy derived for the $1S_{3/2}$–$2P_{3/2}$ transition, $18.5\,\text{meV}$, obtained for a $150\,\text{Å}$ wide well in these resonant luminescence measurements is in excellent agreement with re-

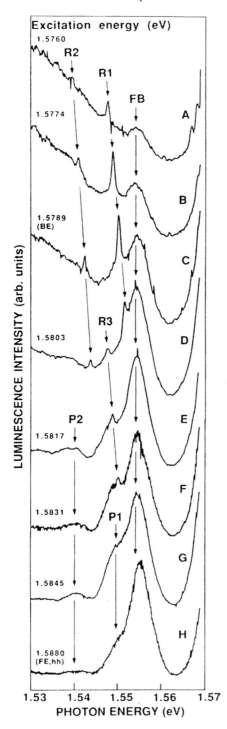

Fig. 6.10. A synopsis of selective photoluminescence spectra for a 69-Å wide quantum well with excitation energies ranging from 1.576 to 1.588 eV, where 1.588 eV corresponds to the energy of the heavy hole state of the free exciton as observed in the photoluminescence excitation spectrum, while 1.579 eV corresponds to the position of the acceptor bound exciton in the photoluminescence spectrum. The R1, R2 and R3 lines are interpreted as acceptor transitions due to electronic Raman scattering, while the P1 and P2 lines are related to the two-hole-transitions corresponding to the same acceptor transitions

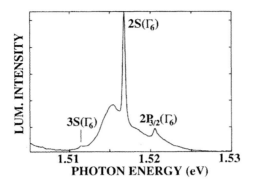

Fig. 6.11. Selective photoluminescence spectrum of a 140 Å wide quantum well doped to a level of $5 \times 10^{16}\,\mathrm{cm}^{-3}$ in the central 28 Å. The selective spectrum is measured at 1.6 K with the excitation energy resonant with the acceptor bound exciton. The two-hole-transition satellite corresponding to the $1S_{3/2}(\Gamma_6)$–$2S_{3/2}(\Gamma_6)$ acceptor transition, denoted as $2S_{3/2}(\Gamma_6)$ in the figure, dominates this spectrum, but a weak feature due to the transition to the next S-like state, the $3S_{3/2}(\Gamma_6)$ state, is also observed. In addition, the two-hole-transition peak originating from the parity-forbidden $1S_{3/2}(\Gamma_6)$–$2P_{3/2}$ acceptor transition is monitored

sults achieved in far infrared absorption spectroscopy (where transitions to P-like excited states dominate the spectra) 18.6 meV, obtained for a similar quantum well [49].

There is another essential aspect associated with the observation of the two hole transitions satellites in resonant photoluminescence: It opens the possibility to monitor the acceptor bound exciton in photoluminescence excitation spectra, which otherwise is difficult for the case of quantum wells. Since the two hole transitions can be regarded as a generalized form of the bound exciton recombination, as described above, an enhancement of the acceptor bound exciton intensity in the photoluminescence excitation spectrum is expected, if the two hole transitions satellite is detected. This fact is illustrated in the synopsis of photoluminescence excitation spectra in Fig. 6.12 for the same 140 Å wide quantum well as used in Fig. 6.11. When the detection energy is resonant with or close to any of the two hole transitions peaks, the bound exciton peak appears in the photoluminescence excitation spectrum in addition to the usually predominant light and heavy hole states of the free exciton, FE^{lh} and FE^{hh} (Fig. 6.12). In fact, similarly to bulk, the appearance of the bound exciton peak in the photoluminescence excitation spectrum proves the interpretation of the usually weak satellite feature as being due to the two hole transitions satellite.

The fact that the acceptor bound exciton can be monitored in photoluminescence excitation spectra by detecting the satellite, can for instance provide essential information on the acceptor ground state. Such an example is illustrated in Fig. 6.12. When the predominant 2S related satellite is

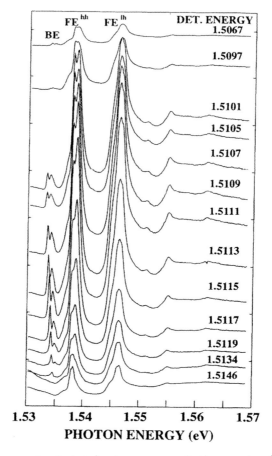

PHOTON ENERGY (eV)

Fig. 6.12. A synopsis of photoluminescence excitation spectra for an acceptor doped $(3 \times 10^{16}\,\text{cm}^{-3})$ $150\,\text{Å}$ wide quantum well. The detection has been close to or resonant with the strongest two-hole-transition satellite (at 1.5113 eV) corresponding to the $1S_{3/2}(\Gamma_6)$–$2S_{3/2}(\Gamma_6)$ acceptor transition. In addition to the dominating heavy hole and light hole states of the free exciton, a doublet structure due to the acceptor bound exciton is observed. This doublet structure exhibits a pronounced thermalization behavior at increasing temperatures, which implies that the splitting originates from the initial state, the acceptor ground state, in the bound exciton excitation process. Accordingly, the doublet structures is interpreted as being due to the splitting between the heavy and light hole state of the acceptor 1S ground state

detected, several peaks in the bound exciton range can be monitored with a predominant doublet structure related to the acceptor bound exciton peak (denoted A_1 and A_2 in Fig. 6.13). As can be seen in Fig. 6.13, the splitting observed between the two components is small and measures just to a fraction of a meV (0.55 meV for the $150\,\text{Å}$ wide quantum well to increase to $0.65\,\text{meV}$ for a $100\,\text{Å}$ quantum well). The two components have been inter-

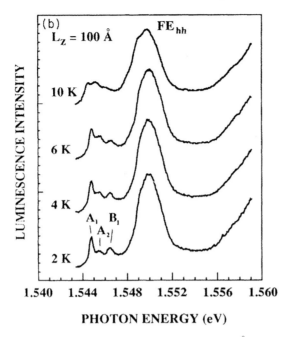

Fig. 6.13. Photoluminescence excitation spectra of a 100 Å wide acceptor doped quantum well detecting the 2S two-hole-transition satellite, measured at some different temperatures to illustrate the different thermalization behaviors of the lines

preted as being due to the splitting between the $1S(\Gamma_6)$ and $1S(\Gamma_7)$ states of the acceptor ground state [157]. These states originate in essence from the heavy- and light-hole ground states of the confined acceptor, although there is a considerable mixture of the states [25]. This interpretation of the $1S(\Gamma_6)$ and $1S(\Gamma_7)$ states is based on polarization, temperature and magnetic field dependent photoluminescence excitation measurements [157].

Since the resonant photoluminescence and photoluminescence excitation techniques are in principle, "reversible" processes, one could expect to make a similar observation of the heavy- and light-hole splitting of the confined acceptor in a resonant photoluminescence spectrum. This behavior is illustrated in Fig. 6.14, in which any of the excitons bound at the heavy hole $1S(\Gamma_6)$ or the light hole $1S(\Gamma_7)$ ground state of the confined acceptor is resonantly excited. In the former case, the satellite peak corresponding to the $1S(\Gamma_6)$–$2S(\Gamma_6)$ transition dominates the satellite spectrum, while a second two hole transition satellite appears on the high energy side of the first peak, when the exciton bound at the $1S(\Gamma_7)$ state is resonantly excited. These satellites are in a natural way interpreted as the corresponding first excited heavy hole and light hole states of the confined acceptor, $2S(\Gamma_6)$ and $2S(\Gamma_7)$, respectively. Also the satellites related to the next S-like state, $3S(\Gamma_6)$ and $3S(\Gamma_7)$, can be observed in a similar way, although at a significantly lower intensity

Table 6.1. The energy positions for different acceptor states relatively the acceptor 1S(Γ_6) ground state in quantum wells of varying well widths estimated from the energy positions of the two-hole transition satellites in selective photoluminescence spectra and compared with theoretical predictions by Fraizoli and Pasquarello [25]. The experimental results for the Be acceptor in bulk GaAs are derived from J.C. Garcia et al. [159]

	50 Å		69 Å		94 Å		150 Å		GaAs
	Expt.	Theory	Expt.	Theory	Expt	Theory	Expt.	Theory	Expt.
1S$_{3/2}$(Γ_7)					0.65±0.10	1.0	0.55±0.10	0.5	
2P$_{3/2}$					19.9±0.2	19.6	18.5±0.2^a	17.0	16.7
2S$_{3/2}$(Γ_6)	31±1	30.2	28.5±0.5	27.1	24.4±0.2	14.4	22.4±0.1	21.9	19.7
2S$_{3/2}$(Γ_7)							25.0±0.1	24.4	
3S$_{3/2}$(Γ_6)			37.0±0.5	33.8	30.2±0.3	30.7	27.7±0.1	27.0	23.7

^{a} L$_z$=140 Å

level. In conclusion, it means that the satellite spectroscopy, performed either in the resonant luminescence mode or in the photoluminescence excitation mode, has provided information on S-like, but also on P-like excited states up to $n = 3$ for both heavy hole, Γ_6-like and light hole Γ_7-like states. A summary of the so-derived acceptor transition energies showing the dependence of the acceptor levels on the quantum well width is given in Table 6.1. The corresponding values for a bulk acceptor are enclosed as a reference.

However, further information can be gained from the photoluminescence excitation measurements, when performed at elevated temperatures. Upon detection of the 2S related satellite and increasing the temperature, the intensity distribution between the bound exciton associated components observed will vary (Fig. 6.13). For instance, the two low-energy components denoted A$_1$ and A$_2$ in Fig. 6.13, separated by 0.55 ± 0.05 meV exhibit a pronounced thermalization behavior: The fact that the A$_2$/A$_1$ intensity ratio increases with increasing temperature, proves that the splitting originates from the initial acceptor ground state. Hence, the A$_1$–A$_2$ splitting is consistent with the splitting of the acceptor heavy hole-like 1S(Γ_6) and the light hole-like 1S(Γ_7) state. The two strong peaks, denoted A$_1$ and B$_1$ in Fig. 6.13, separated by 1.1 meV, exhibit a different behavior with increasing temperature with an intensity ratio, which remains about the same at elevated temperatures. Such a temperature dependence is consistent with two states originating from the final state in absorption, i.e. from the excited state, the bound exciton state. This A$_1$–B$_1$ splitting has been explained in terms of the splitting between the $J = 5/2$ and $J = 3/2$ exciton states [157]. The "missing" feature corresponding to B$_2$, i.e. another transition involving the acceptor light hole-like 1S(Γ_7) state (see Fig. 6.9), can be observed at elevated temperatures, though weak and broad, in the tail of the heavy hole state of the free exciton peak.

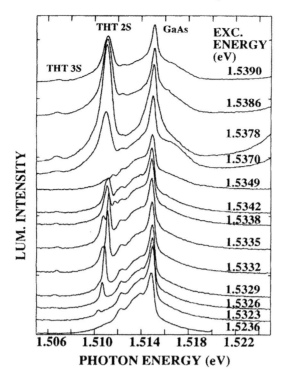

Fig. 6.14. A synopsis of selective photoluminescence spectra for an acceptor doped $(3 \times 10^{16}\,\mathrm{cm}^{-3})$ 150 /AA wide quantum well (the same well as that used in Fig. 6.7 and used in Fig. 6.12), measured at 1.6 K with varying excitation energy in the exciton range. Upon excitation resonant with the acceptor bound exciton, a splitting of the two-hole satellite is resolved. The photoluminescence features at higher energies are due to GaAs emissions as shown in the bottom spectrum with the excitation energy well below the quantum well transition energy

The interpretation of the A_1 and A_2 peaks as being due to the acceptor heavy hole-like $1S(\Gamma_6)$ and the light hole-like $1S(\Gamma_7)$ states, respectively, was confirmed by polarized photoluminescence excitation measurements. Circularly polarized light from a photoelastic modulator was used to excite the sample. The differential signal between the left-hand and right-hand circularly polarized recombination light from the sample was monitored. Primarily, the behavior of the free exciton is focussed. For a polarized photoluminescence excitation spectrum for a 150 Å wide quantum well, the negative differential output signal is obvious for the light hole free exciton state, while the heavy hole free exciton exhibits an analogous positive differential signal, as illustrated in Fig. 6.15. For reference, the corresponding unpolarized photoluminescence excitation reference spectrum is exposed.

Next, the behavior of the acceptor bound exciton in polarized photoluminescence excitation spectra will be focussed. When detecting the 2S related

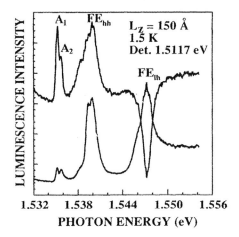

Fig. 6.15. Polarized photoluminescence excitation spectrum (*the upper spectrum*) of the 150 Å wide quantum well together with an unpolarized reference spectrum (*the lower spectrum*). The difference in the luminescence intensity from the sample is detected for excitation by left-hand and right-hand circularly polarized excitation light, respectively, in the polarized spectrum

satellite, the different bound exciton peaks appear in the spectrum, but all of them exhibiting a positive differential signal in contrast to the case of the free exciton as demonstrated above. However, the intensity ratio deviates significantly for the A_1 and A_2 bound exciton peaks relatively the corresponding free exciton peaks. By comparing the intensity ratio for the A_1 peak versus the A_2 peak in the unpolarized excitation reference spectrum with the differential output signal in the polarized luminescence excitation spectrum, one can conclude that although the A_2 peak exhibits a positive signal (Fig. 6.15), it is significantly reduced compared to its intensity in the unpolarized reference spectrum. This situation reflects most likely the condition described above, namely that the acceptor states are strongly mixed and an intensity ratio different from the corresponding heavy and light hole states of the free exciton is consistent with expectations. In fact, the observed A_1/A_2 intensity ratio gives a hint about the heavy/light hole mixing rate for the acceptor states.

6.2.8 The Dependence of the Binding Energy on the Position in the Well

As an acceptor is moved away from the center of the quantum well towards the interface, the charge density is also shifted towards the well edge, but considerably less than the impurity. For an on-edge acceptor, the charge distribution is asymmetric, but the charge density is still mainly confined in the quantum well and the wave functions essentially keep their characters [111]. As the acceptor is moved further out in the barrier, the charge distribution

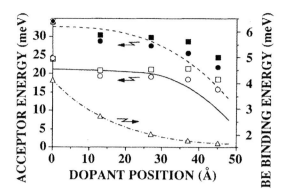

Fig. 6.16. The experimentally derived binding energies of the acceptor bound exciton, represented by open triangles vs. the dopant position z_0 in the well. The *dash-dotted line* is a fit to the experimental data. The upper set off data corresponds to the 1s–2p acceptor transition energies and the total acceptor binding energies vs. the dopant position z_0 in the well. The *squares* correspond to the two-hole-transition data and the *circles* are from resonant Raman scattering data

is still confined in the quantum well. This means that while the positively charged acceptor is moved away towards the interface, the hole distribution is still confined close to the center of the quantum well. Consequently, the attractive potential energy and hence the binding energy is reduced as the acceptor is moved from the center of the quantum well towards the interface. When the acceptor is moved further out of the well and into the barrier, the charge distribution becomes in fact more symmetric in the central part of the quantum well again with a further reduced binding energy as a result. A. Pasquarello et al. [111] explains this effect as due to the fact that only the tail of the Coulomic potential is effective and instead the confining barriers of the quantum well predominate.

The dependence of the binding energy of an acceptor bound hole, confined in a quantum well, on the impurity position was originally calculated by Bastard [21]. In his model, a hydrogenic impurity and infinite barrier height was assumed. Later on, more realistic calculations with finite barriers were reported by Masselink et al. [108, 158]. Pasquarello et al. [111] calculated also the binding energies of the excited acceptor states. R.C. Miller [160] investigated the position dependence of the acceptor by means of luminescence spectroscopy. The luminescence spectra of two quantum well structures doped with Be acceptors in a layer corresponding to 25% of the well width either in the center of the well or at the interface. It was demonstrated from the energies for the free electron recombining with an acceptor bound hole, i.e., a free-to-bound transition, e-Be^0, that the acceptor binding energy clearly decreases as the acceptor is moved away from the center of the well towards the interface. For instance the total binding energies, estimated from these

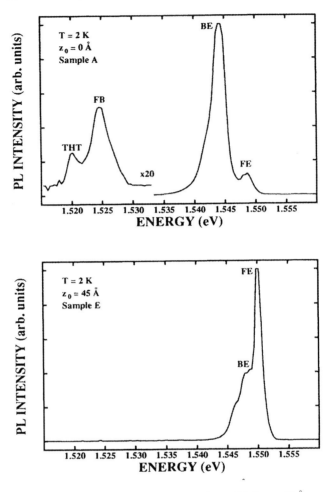

Fig. 6.17. Photoluminescence spectra measured at 2 K for a 96 Å quantum well with (**a**) acceptors doped in the center of the well ($z_0 = 0$) and (**b**) acceptor doped close to the interface of the well ($z_0 = 45$ Å). The energy separation between the bound exciton and the free exciton decreases drastically as the acceptors are moved towards the interface (compare with Fig. 6.16). The acceptor doping level is about 1×10^{17} cm^{-3}

transitions, is 34 meV for acceptors at the center of a 100 Å wide well, while it has decreased to 18 meV for acceptors at the well edge.

The dependence of the acceptor energies on the spatial position of the acceptor within the quantum well was studied by G.C. Rune et al. [139] by means of satellite spectroscopy. Both resonant Raman scattering and Two Hole transitions of the acceptor bound exciton recombination (as described above in Sect. 6.2.7) were monitored for a series of Be δ-doped samples with different positions of a thin dopant layer (approximately 2–3 monolayers) from the center out to the edge of the well. The result of this systematic study on the energy of the 1s–2s transition as a function of the dopant position in a 100 Å wide quantum well is given in Fig. 6.16. For the same set of samples, the dependence of the binding energy of the acceptor bound exciton on the dopant position is presented.

Already in a non-selective luminescence spectrum (Fig. 6.17), it is clearly illustrated that the bound exciton binding energy, i.e. the energy difference between the free exciton and the bound exciton decreases drastically as the acceptor is moved away from the center; from 4.2 meV for center bound exciton to 1.6 meV for an exciton bound to an acceptor close to the interface in a 100 Å wide quantum well. This fact can be explained in terms similar to what has been argued above for the acceptor binding energy as the acceptors are displaced from the center of the well, i.e., the bound exciton wave function will still be confined close to the center of the quantum well although the acceptor is moved away towards the interface. Accordingly, the overlap between the acceptor wave function and the bound exciton wave function decreases with a reduced attractive potential energy and bound exciton binding energy as result.

In a luminescence spectrum with excitation close to or resonant with the acceptor bound exciton, satellites originating from both resonant Raman scattering and Two Hole transitions of the acceptor bound exciton can be monitored. A typical series of spectra for an off-center doped sample (with a dopant δ-layer positioned 10–16 Å away from the center of a 100 Å wide quantum well) is shown in Fig. 6.18. Both satellites corresponding to the resonant Raman scattering and the two hole transitions are observed in this spectrum. Similar selection rules apply for these two processes: Transitions between states of the same parity are favoured, i.e. mainly from $1S_{3/2}$ ground state to s-like excited states [161]. The predominant acceptor transition observed for both these transition mechanisms is the 1s–2s transition. As seen in the spectrum, the energy separation between the excitation energy and the sharp peak denoted R remains constant, as characteristic for Raman scattering peaks. The intensity of satellites associated with acceptors in the center of the quantum well is enhanced by the density of states per energy, $g_L(E) = (2/L)(dz_0/dE)$, which increases towards the center to become infinite at the center, since dE/dz vanishes at the center [21, 111]. This means that also a minor concentration of non-intentional acceptors at the center of the quantum well will give rise to a significant intensity level for satellites associated with acceptors in the center of the quantum well. Accordingly, not only satellites originating from acceptors in the dopant δ-layer central acceptors are observed but also satellites with an origin from acceptors in the center of the well.

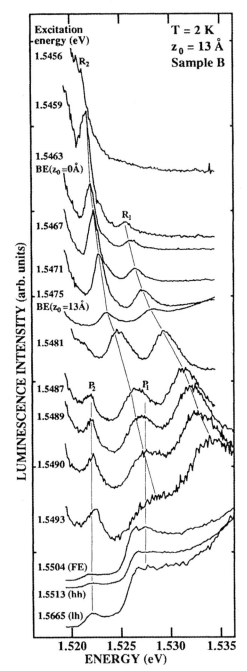

Fig. 6.18. Selective photoluminescence spectra of an acceptor doped quantum well with the acceptor dopant layer displaced 13 Å from the the center of the well, i.e. $z_0 = 13$ Å. The notations (hh) and (lh) denote resonant excitation with the heavy-hole and light-hole states of the free exciton (FE) as observed in photoluminescence excitation spectrum. The notations FE, BE($z_0 = 13$ Å) and BE($z_0 = 0$ Å) denote resonant excitation with the free exciton and the exciton bound to acceptors at $z_0 = 13$ Å and at the center of the well ($z_0 = 13$ Å), respectively

6.2.9 Magneto-optical Properties

A magnetic field applied on a hydrogenic system results in the Zeeman effect. The Hamiltonian with isotropic single band model including the Zeeman effect can be expressed as [108]

$$H = \frac{\hbar^2 k^2}{2m} + \frac{eB}{2mc}(L_z + 2S_z) + \frac{e^2 B^2}{8mc^2}(x^2 + y^2) \qquad (6.11)$$

with the applied magnetic field, B, in the z direction and the two last terms representing the linear and quadratic Zeeman components, respectively. The situation is more complicate in the case of acceptors confined in a quantum well structure, a four bands effective mass model has to be used, as discussed in Sect. 6.1. In the presence of a magnetic field, the twofold degeneracy of the acceptor heavy hole ground state is expected to be lifted. The linear Zeeman splitting, ΔE, for the acceptor heavy hole ground state with an effective g-value, g_{eff}, is derived from

$$\Delta E = g_{eff} B \mu_B \qquad (6.12)$$

where B is the applied magnetic field and μ_B is the Bohr magneton.

When investigating the magnetic field perturbation on a confined acceptor by means of luminescence spectroscopy, it is usually performed via the exciton bound at the acceptor. The magnetic field dependence will below be exemplified with partly the recombination of acceptor bound exciton and partly the transitions between the acceptor ground state and excited states as derived from satellite spectroscopy.

The initial state of the luminescence emission, the acceptor bound exciton state, is a three particle state with two holes and one electron. If this is treated similarly to the bound exciton in the bulk case, the j–j coupling between the two bound holes and the single electron involved in the bound exciton system gives rise to three bound exciton states with $J = 1/2, 3/2$ and $5/2$. However, also the quantum well confinement effect has to be considered in a proper treatment of the bound exciton in the 2D case. The quantum well confinement effect on the acceptor bound exciton is introduced in the calculations via the axial crystal field perturbation. The interaction between the three electronic particles has been calculated by using an effective perturbation spin-Hamiltonian, which consists of three terms describing the hole-electron and hole-hole interaction and the crystal field effect,

$$H = H_{h-e} + H_{h-h} + H_c , \qquad (6.13)$$

where

$$H_{h-e} = -A(\boldsymbol{J_1} + \boldsymbol{J_2}) \cdot \boldsymbol{S} ,$$
$$H_{h-h} = -B\boldsymbol{J_1} \cdot \boldsymbol{J_2} ,$$
$$H_c = -D\left[J_{1z}^2 + J_{2z}^2 - 1/3J_1(J_1 + 1) - 1/3J_2(J_2 + 1)\right], \qquad (6.14)$$

where J_1 and J_2 are the angular momenta of the two bound holes and S the angular momentum of the electron. The D parameter is essential for describing the axial symmetry. The A and B parameters describe the electron-hole and hole-hole interaction, respectively. For the case of bulk GaAs, we know that j–j coupling of the two bound holes gives rise to two states, $J = 0$ and $J = 2$, of which the latter $J = 2$ state is at lowest energy for shallow acceptors in GaAs. The wavefunction $(\mid J = 0\rangle + (\mid J = 2\rangle) \otimes (\mid S = 1/2\rangle$ can then be used as the unperturbed eigenstate. The perturbed eigen-energies and wavefunctions are derived from the secular equation

$$\det \parallel H_{ij} - E\,\delta_{ij} \parallel = 0 \tag{6.15}$$

where $H_{ij} = \langle \phi_j \mid H \mid \phi_j \rangle$.

The calculatations of the electronic structure was based on twelve unperturbed wavefunctions, constructed from $(\mid J = 0\rangle + (\mid J = 2\rangle) \otimes (\mid S = 1/2\rangle)$ with H defined according to (6.13) and ϕ_i ($i = 1, 2, \ldots 12$). Without confinement, i.e., corresponding to the bulk case with $D = 0$, the two lowest acceptor bound exciton states are the $J = 5/2$ and $J = 3/2$ states. When the crystal field perturbation is applied, i.e. for the quantum well case with $D > 0$, these states are further split. Due to the strong mixing between the $J = 0$ and $J = 2$, the $J = 3/2$ state splits into two doublet states, $m = \pm 1/2$ and $m = \pm 3/2$, while the $J = 5/2$ state at lowest energy splits into one doublet state, $m = \pm 1/2$, and one fourfold degenerate state, $m = \pm 3/2, \pm 5/2$ (see Fig. 6.9) by the axial H_c perturbation.

This means that the bound exciton emission detected in photoluminescence at low temperatures occurs between the initial $J = 5/2$ acceptor bound exciton state and the final acceptor $1S(\Gamma_6)$ ground state. The calculated dependence of the energy positions as well as the allowed transition energies on the magnetic field is shown in Fig. 6.19. The *solid lines* show the σ polarized transitions, while the *dashed lines* show the π polarized transitions, which are forbidden in the Faraday configuration. The resulting splitting for the allowed transitions between the bound exciton state and the $1S(\Gamma_6)$ ground state of the acceptor as a function of the magnetic field is shown in the *top part* of Fig. 6.19.

The satellite peaks related to the acceptor $1S(\Gamma_6)$–$2S(\Gamma_6)$ transitions shift towards higher energy with increasing magnetic field like the acceptor bound excitons, but at a slightly slower rate than the principal bound excitons. This means that the energy separation between the acceptor bound excitons and the two hole transition satellites, i.e., the $1S(\Gamma_6)$–$2S(\Gamma_6)$ transition energy, increases with increasing magnetic field. This observation applies to various well widths investigated, although to different extent. The dependence of the acceptor $1S(\Gamma_6)$–$2S(\Gamma_6)$ transition energy on the applied magnetic field for three different well widths is shown in Fig. 6.20. The energy levels for the acceptor $1S(\Gamma_6)$ ground state and the excited $2S(\Gamma_6)$ state in the presence of a magnetic field are schematically illustrated in Fig. 6.19 for the case of acceptors in the center of a 150 Å wide quantum well. The predicted splitting

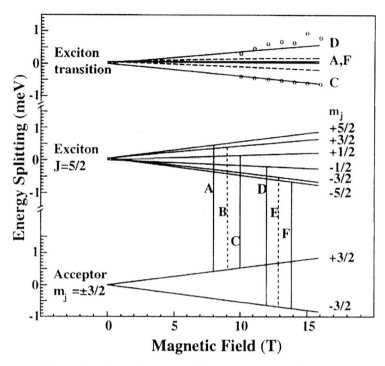

Fig. 6.19. The predicted development of the acceptor $1S_{3/2}(\Gamma_6)$ ground state and the $J = 5/2$ bound exciton states, with increasing magnetic field, for a 150 Å quantum well is shown in the *lower part* of this figure. The allowed transitions between these components (denoted A–F) are given by *solid lines* corresponding to the σ-polarized transitions and *dashed lines* for the π-polarized transitions (forbidden in the Faraday configuration). The resulting splitting for the allowed transitions between the acceptor bound exciton states and the $1S_{3/2}(\Gamma_6)$ ground state of the acceptor as a function of the magnetic field is shown in the *top part* of this figure. In order to compare with experimental results, the splitting of the acceptor bound exciton observed in photoluminescence measurements is plotted relative to the low energy branch (denoted C)

between the Kramers' doublet states, i.e. the spacing between the $m_j = \pm 3/2$ components, is significantly larger for the $1S(\Gamma_6)$ ground state than for the excited $2S(\Gamma_6)$ state. Due to the two-fold degeneracy of these states, there are four possible transitions between between the acceptor $1S(\Gamma_6)$ ground state and the excited $2S(\Gamma_6)$ state (denoted A, B, C and D in Fig. 6.19), of which two transitions, $m_j = \pm 3/2 \longleftrightarrow m_j = \pm 3/2$, involve spin-flip [124]. Due to the small splitting between the $m_j = \pm 3/2$ components of the $2S(\Gamma_6)$ excited state relatively the corresponding splitting of the $1S(\Gamma_6)$ ground state, there are two major branches (within the eperimental resolution limits). A fitting procedure of the experimental data for the high energy branch for the

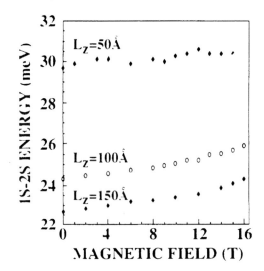

Fig. 6.20. The experimentally determined energy separation between the acceptor 1S(Γ_6) ground state and the excited 2S(Γ_6) state as a function of an applied magnetic field for three different quantum well widths. The experimental estimates are based on the results achieved from electronic Raman scattering and the two-hole-transitions experiments

acceptor 1S(Γ_6)–2S(Γ_6) transition energy with the theoretical predictions, results in a value of the Luttinger parameter of $\kappa = 1.2$ [117, 162].

In order to evaluate the magnetic field dependence of the acceptor bound exciton emission, one has to take into account that the emission corresponds to a transition between the acceptor bound exciton as initial state and the confined acceptor states as the final state. This fact constitutes a considerable complication, when analysing the effect of the magnetic field on the acceptor, since the electronic structure of the bound exciton state, involving three electronic particles is quite complex. The magnetic field dependence of a bound exciton recombination will be examplified by a 150 Å wide quantum well doped with Be in the central 30 Å of the quantum well.

First of all, when such a quantum well is exposed to a magnetic field, one observes a spectral blue shift of both the dominating free exciton and the acceptor bound excitons due to the diamagnetic shift. This diamagnetic shift, the quadratic Zeeman effect, can be derived from the second order perturbation theory of an hydrogenic impurity [163]

$$\Delta E_{\text{dia}} = \frac{1}{2} \left(\frac{e\,B}{c} \right)^2 \frac{a_0^2}{2\mu_0} \tag{6.16}$$

where $\mu_0 = m_0/\gamma_1$ (γ_1 is the Luttinger parameter) and a_0 is the Bohr radius of the ground state wave function of the acceptor defined as $a_0 = (\epsilon\,\hbar^2/e^2\,\mu_0)$. The diamagnetic shift rate observed for the acceptor bound exciton recom-

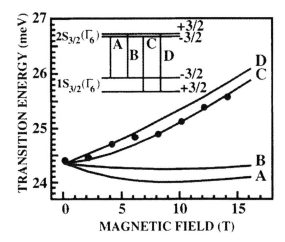

Fig. 6.21. The inset shows schematically the splitting of the $m_j = \pm 3/2$ components for the $1S_{3/2}(\Gamma_6)$ ground state and the $2S_{3/2}(\Gamma_6)$ excited state of the confined acceptor in the presence of a magnetic field. The possible transitions between the acceptor $1S_{3/2}(\Gamma_6)$ ground state and $2S_{3/2}(\Gamma_6)$ excited state are denoted by A-D. The lower figure shows the calculated transition energies for the case of an acceptor in the center of a 100 Å wide quantum well. The experimental results are obtained from the resonant Raman scattering and two-hole-transition experiments and are given as circles in the figure

bination decreases in a quantum well compared to the bulk case (e.g. from 2.8×10^{-2} meV/T^2 for the bound exciton in GaAs bulk to 2.1×10^{-2} meV/T^2 for the exciton in a 150 Å wide quantum well [117,162]). This effect can be expected since the reduction of the diamagnetic shift corresponds to a compression of the exciton wave function. The observed decrease of the diamagnetic shift rate corresponds to a reduction of the hydrogenic orbital radius by 13%, when going from bulk GaAs to a 150 Å wide GaAs/AlGaAs quantum well.

Further, for a sufficiently high magnetic field, the bound exciton peak is found to split into two components in a photoluminescence spectrum [157]. The interpretation of this linear Zeeman splitting is not straight forward, since not only the splitting of the acceptor ground states ($m_j = \pm 3/2$ for the $1S(\Gamma_6)$ state and $m_j = \pm 1/2$ for the $1S(\Gamma_7)$ light hole-state), but also of the excited acceptor bound exciton states in the presence of a magnetic field. Consequently, the splitting observed of the acceptor bound exciton emission is the combined effect of the splitting in the acceptor ground state and the acceptor bound exciton state and the g-values for the electron and hole individually can not be directly determined from these spectra. It is rather the effective g-value, g_{eff}, for the bound exciton emission, which can be evaluated. A striking enhancement of these effective g-values are found, when going from the case of an acceptor bound exciton in bulk to the corresponding bound

exciton in a quantum well (e.g from $g_{\mathrm{eff}} = 0.66$ for the bound exciton in bulk
GaAs [117, 162] to $g_{\mathrm{eff}} = 1.89$ for a 100 Å wide quantum well).

In order to evaluate the g-values individually for the electron and hole,
respectively, the electronic structure of the acceptor ground state and the
excited acceptor bound exciton state have to be separately treated. For the
acceptor $1S(\Gamma_6)$ ground state, the linear magnetic field splitting between the
acceptor $\pm \mid J_z \mid$ components can be described by

$$\Delta E = g_{J_z} \, 2 \mid J_z \mid B \, \mu_B \qquad (6.17)$$

where J_z is the magnetic angular momentum in the direction of the magnetic
field and g_{J_z} is the related g-value. The so-derived calculated g-values for the
confined acceptor holes [117] are in good agreement with the experimental
data available [124–126]. The splitting for the acceptor $1S(\Gamma_6)$ ground state,
i.e. the final state in the photoluminescence emission, in a 100 Å wide quantum well for a g-value of $g_{3/2} = 0.61$ and $g_{1/2} = 0.35$ [117, 162] is shown in
Fig. 6.21. The splitting observed for the bound exciton recombination originates mainly from the splitting in the final state of the emission, i.e. between
the $m_J = \pm 3/2$ states of the acceptor. The splitting in the initial state, the
bound exciton state, is assumed to be small relatively the ground state splitting (see Fig. 6.19). Accordingly, the evaluated effective g-value, g_{eff}, from
the observed splitting is dominated by the acceptor hole splitting. For the
quantum wells used, an effective g-value, g_{eff}, of $g_{\mathrm{eff}} = 1.89$ is estimated for
the acceptor bound exciton in a 100 Å wide quantum well, to be compared
with a g-value of $g_{\mathrm{eff}} = 0.66$ for the corresponding exciton bound at a shallow
acceptor in bulk GaAs.

6.2.10 Strain Effects on the Electronic Structures of Acceptors

According to the theoretical calculations, both the quantum well confinement and biaxial deformation potential have a strong effect on the electronic
structure of the acceptor [115]. In general, a strong quantum well potential leads a large binding energy of the confined shallow acceptors, while
a biaxial potential reduces the binding energy of acceptor. In addition, the
quantum well potential and biaxial potential also have different effects on the
energy separations between the acceptor ground state and the excited states.
Loehr et al. [109] made a theoretical and experimental study on the ground
state of acceptors confined in strained quantum well systems. They found
that the binding energy of the acceptor 1s state decreases with increasing
built-in strain. The detailed electronic structure of the acceptors confined in
$In_xGa_{1-x}As/AlGaAs$ quantum well structures were investigated as a function
of the In-concentration in the quantum well structures, where the acceptors
were located in the central region of the well layers.

The spectra from a set of samples with deffirent In-concentrations were
measured. The luminescence spectra with above-band-gap excitation together

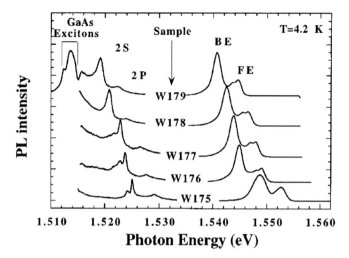

Fig. 6.22. Photoluminescence and selective photoluminescence for five different samples with different In-concentration in an $In_x Ga_{1-x}As/Al_{0.3}Ga_{0.7}As$ quantum well structure, measured at 4.2 K. The luminescence spectra were recorded with excitation wavelength of 5145 Å, while the selective luminescence was measured with an excitation energy close to resonance with the acceptor bound exciton. The excitation density was about $1.1\,W/cm^2$

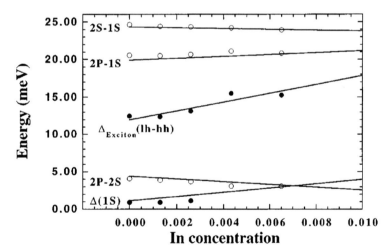

Fig. 6.23. Transition energies for the confined acceptor transition as the In-concentration is varied in an $In_x Ga_{1-x}As/Al_{0.3}Ga_{0.7}As$ quantum well. The filled and open circles represent the experimental results based on Resonant Raman satellites, while the solid lines correspond to the theoretically calculated data

with the selective photoluminescence spectra with the excitation energy close to resonant of the acceptor bound exciton are shown in Fig. 6.22 for five samples with various In-concentrations. The peak appearing between the bound exciton and the free exciton transitions in some photoluminescence spectra is related to the biexciton. The excitation density is of $1.1\,\mathrm{W/cm^2}$. The red shift of the bound exciton and the free exciton transition energies with increasing In-concentration is due to the narrowing of the InGaAs band gap. The selective photoluminescence spectra at the energy around $1.52\,\mathrm{eV}$ show a similar feature for all samples. However, the 2S and 2P satellites clearly illustrate an increase of the line-width with increasing In-concentration. The splitting of the acceptor $2S_{3/2}(\Gamma_6)$ state is hardly resolved when the In-concentration in the structures is larger than 0.3%. The increase of the line-width of the Resonant Raman satellites is due to fluctuation of the In-concentration in the structures. Such an effect is well-known in MBE grown InGaAs materials. The experimental data and theoretical calculated results are summarized in Fig. 6.23. With increasing In-concentration x, the separation between the light and heavy hole states increases due to the built-in strain effects. The solid lines in Fig. 6.23 correspond to the calculated results and the dots have been deduced from the experiments [164]. Both the tendency and absolute energy values of the acceptor states with varying In-concentration show an excellent agreement between the experimental and the calculated results. It should be pointed out that according to the theoretical calculations, the change of the acceptor electronic structure, in the range of In-concentrations used here, is mainly due to the variation of the biaxial potential.

6.2.11 Dynamics

The effect of the Coulombic electron-hole interaction is reflected by the radiative kinetics of the free excitons. Their interplay with impurities is similarly revealed by corresponding studies of the bound excitons. The excitons in quasi two-dimensional structures exhibit different properties as compared with corresponding three-dimensional semiconductors due to the confinement effects on the electrons and holes. An apparent confinement effect is the increased binding energy of the excitons. Also the probability for a recombination, as reflected by the luminescence decay of the exciton is affected by the confinement. The oscillator strength per unit cell of a quasi-2D free exciton is given by [165]

$$f_x^{2D} = \Omega f_0 \mid \phi_{1s}^{2D}(0) \mid^2 / L_z \qquad (6.18)$$

where Ω is the volume of the unit cell, f_0 is the dipole matrix element, which connects Bloch states in the conduction and the valence bands and ϕ_{1s}^{2D} is the 2D-like wave function for the 1s state of the electron-hole pair. The probability for the electron and hole to be in the same unit cell is given by

$$\mid \phi_{1s}^{2D}(0) \mid^2 = \frac{2}{\pi a_0^2} = \frac{2}{A_x} \qquad (6.19)$$

where A_x is the area of the 2D exciton with the Bohr radius a_0. The lifetime is correlated with the oscillator strength,

$$\tau \propto \frac{1}{f_x^{2D}} \tag{6.20}$$

A linear decrease of the exciton decay time with decreasing quantum well width is predicted from these theoretical considerations.

Acceptors Confined in Isolated Quantum Well Structures

If the barrier between the quantum well is thick enough, such as 100 Å wide AlGaAs in GaAs/AlGaAs structures, a multiple-quantum well structures can be treated as an isolated quantum well. The decay time can still be estimated according to (6.18). This predicted linear dependence has been experimentally demonstrated [165–167]. However, some controversy exists, whether the exciton recombination originates from laterally localized excitons or from excitons with mobility in areas larger than the exciton Bohr radius. For the "free" exciton, it has been concluded that potential fluctuations due to variations in the well width play an important role in the capture and kinetic behaviour of the excitons [168–171].

In order to maintain momentum conservation, preferentially $K = 0$ excitons recombine. However, with increasing temperature, a redistribution of the exciton momentum occurs. This fact results in a decreasing fraction of excitons with $K = 0$, which fulfill the conservation rule for radiative recombination [165]. Consequently, the exciton decay times increase with increasing temperature. The same effect has been observed in bulk GaAs [172, 173], although the effect is more pronounced in quantum wells due to the absence of competing non-radiative recombination channels at low temperatures. However, at elevated temperatures, the onset of non-radiative processes makes the effect less obvious.

For the case of acceptor doped quantum wells, the measured decay time of the free exciton is shorter than the corresponding measured exciton decay time in an undoped quantum well (by a factor of approximately 2 for a 150 Å wide well doped to a level of 2×10^{17} cm^{-3} in the central 20% of the well [58]). The measured decay times of the free and acceptor bound excitons recombinations are shown in Fig. 6.24 for some GaAs/AlGaAs quantum structures with different well width. As can be seen in this figure, the measured decay time of the free exciton decreases with increasing well width, but this fact is related to the increasing width of the doping layer, i.e., a higher efficient doping level in wider wells. It can also be concluded from this figure that the measured decay time of the bound exciton is longer than the corresponding decay time for the free exciton at lowest temperature, but is significantly shorter (by about a factor of 2 for a 100 Å wide well) than for the acceptor bound exciton in bulk GaAs.

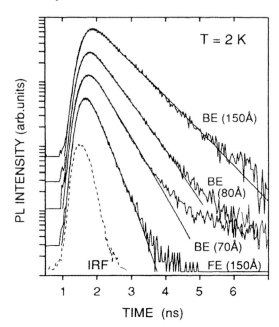

Fig. 6.24. Typical photoluminescence decay curves, measured at 2.0 K with the excitation resonant with the light hole free exciton state. The different curves are from the top: the bound exciton recombination in a quantum well with $L_z = 150$, 80, and 70 Å, respectively; the free exciton recombination in the 150 Å wide quantum well; the lowest dashed curve represents the instrumental response of our detection system. The solid lines corresponding to the best fit according to the equation: $I(t) = \int_0^t F(t_1)e^{-(t-t_1)W}\,dt_1$ are also shown, respectively. All curves are normalized, vertically shifted, and shown in a logrithmic scale

The decay times of the acceptor bound exciton exhibit a weak but significant correlation with the quantum well width. To the best of our knowledge, no theoretical calculations have been performed to predict the kinetic behavior of the acceptor bound exciton in quantum wells. For the donor bound excitons, on the other hand, calculations have been reported [173]. Since the quantization effect should influence the acceptor bound exciton in a similar way as for the donor bound exciton, a comparison with these results are appropriate. Based on these calculations, a decrease of the oscillator strength, corresponding to an increased decay time, was predicted for excitons in quantum wells. Also, a decay time almost independent of the well width was predicted for narrow wells, but for wider wells a width dependence must be included in order to be able to continuously approach the GaAs bulk value of 1.0 ns [174]. With increasing temperature, a redistribution of the momentum for the free exciton will occur as described above. If only excitons within the spectral width $\Delta(T)$ contribute to the transition, there will just be a limited

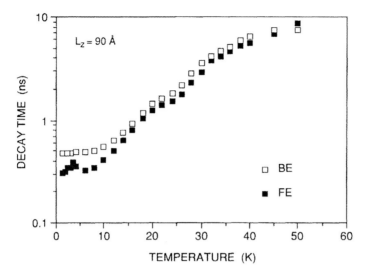

Fig. 6.25. The temperature dependence of the bound exciton and the free exciton decay times for a quantum well with a width of 90 Å. The exciton decay times increase with increasing temperature, as can be seen in the figure, which is due to the reduced fraction of free excitons with $K = 0$, for which the momentum conservation rule required for radiative recombination is accomplished

fraction $\rho(T)$ of free excitons involved in the transition [165].

$$\rho(T) = 1 - \exp(-\frac{\Delta(T)}{kT}) \tag{6.21}$$

The temperature dependence of the observed decay time for the free exciton can be written as

$$\tau(T) = \tau(0)\frac{\Delta(T)}{\rho(T)} \tag{6.22}$$

Due to the resulting reduced fraction of free excitons with $K = 0$, for which the momentum conservation rule required for radiative recombination is accomplished, the exciton decay times increase with increasing temperature as illustrated in Fig. 6.25. In the same figure, also the temperature dependence of the measured decay time for the bound exciton in the same quantum well sample is depicted.

For the bound exciton, on the other hand, a different temperature dependence of the decay time in comparison with the free exciton is expected, since localized particles like a bound exciton, has no momentum distribution. At higher temperatures, however, the free and bound excitons exhibit a similar dynamics response. This fact implies that there is a thermal interaction between these two exciton states, i.e. that a free exciton can be captured at an acceptor to form a bound exciton and inversely a bound exciton can be thermally released from the acceptor into a free exciton state.

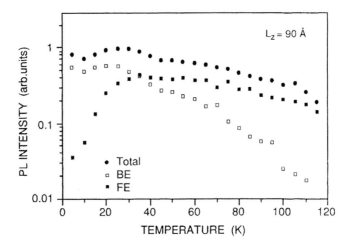

Fig. 6.26. The temperature dependencies of the luminescence intensity for the free exciton and bound exciton recombinations in a doped quantum well with a width of 90 Å. The total luminescence intensity is an integration of the band-gap region in the measured spectra

The observed temperature dependencies of the integrated luminescence intensity for the free exciton and bound exciton are demonstrated in Fig. 6.26, for a sample with a well width of 100 Å. At low temperatures the bound exciton emission dominates, but it is rapidly quenched as the temperature is increased. At the same time the intensity of the free exciton recombination increases so that the total luminescence intensity remains constant up to about 100 K, which can be seen in Fig. 6.26. The quenching of the bound exciton luminescence corresponds to an activation energy of 4 meV, i.e., approximately corresponding to the bound exciton binding energy as observed from the energy separation between the free exciton and the acceptor bound exciton in luminescence spectra.

If the temperature dependencies for the time integrated luminescence intensity and the measured decay times are combined in coupled rate equations

$$\frac{dn_{FE}}{dt} = G(t) - W_{FE}n_{FE} - W_c n_{FE}(N_{A^0} - n_{BE}) + W_R n_{BE}$$

$$\frac{dn_{BE}}{dt} = -W_{BE}n_{BE} - W_c n_{FE}(N_{A^0} - n_{BE}) - W_R n_{BE} \qquad (6.23)$$

for the free and bound excitons, a capture rate, W_c, can be evaluated for the case of low temperatures [58]. W_R corresponds to the thermal release rate of bound excitons from the acceptor into free excitons. The $G(t)$ is the generation rate of the free excitons during the laser excitation and N_{A^0}, n_{FE} and n_{BE} correspond to the concentrations of neutral acceptors, free excitons and bound excitons, respectively. W_{FE} and W_{BE} are the radiative transition rate for the free and bound excitons, respectively.

$$W_{\mathrm{c}} = \frac{I_{\mathrm{BE}}}{I_{\mathrm{FE}}} \left(\frac{W_{\mathrm{FE}}}{N_{\mathrm{A}^0} - n_{\mathrm{BE}}} \right) \tag{6.24}$$

where the I_{BE} and I_{FE} is the time integrated luminescence intensity. For the case of a quantum well with a width of $80\,\text{Å}$, doped with acceptors in the center of the well to a sheet concentration of $1.6 \times 10^{10}\,\mathrm{cm}^{-2}$, a capture rate of $W_{\mathrm{c}} = 0.12\,\mathrm{cm}^2\mathrm{s}^{-1}$ was derived. The capture rate is important for the measured decay time of the free exciton, $\tau_{\mathrm{exp}}^{\mathrm{FE}}$, at low temperatures according to

$$\tau_{\mathrm{exp}}^{\mathrm{FE}} - \frac{1}{W_{\mathrm{exp}}} = \frac{1}{W_{\mathrm{FE}} + W_{\mathrm{c}}(N_{\mathrm{A}^0} - n_{\mathrm{BE}})} \tag{6.25}$$

since the second term in the denominator is introduced due to the presence of acceptors and the excitons bound at these acceptors.

The 2D–3D Transition and the Dynamic Properties of Acceptor Bound Exciton

The transition from a 3D-like impurity to a corresponding 2D-like impurity offers an interesting and exciting transformation from a fundamental physics point of view. As an impurity is subjected to an increasing confining potential, i.e. going from the 3D case towards a 2D-like impurity, the properties of the impurity will change in many aspects as described above. One possibility to follow this transition and the associated properties of the impurity has been demonstrated for a GaAs/AlGaAs quantum well system with a varying barrier width, while the well width is maintained constant [175,176]. For sufficiently thick barriers, the neighbouring wells are electronically isolated, i.e. the electron and hole exhibit 2D character. As the barrier width is decreased, there will be an interaction between neighbouring wells and they become electronically coupled. The structure turns instead into a gradually more 2D-like system. For the extreme case, when the barrier width becomes zero, the system becomes bulk-like. One has accordingly a system, for which one can control the 2D character relative to the 3D character by varying the barrier width.

In this way, the properties of the acceptor and its bound exciton have been investigated as a function of barrier width. There is an overall redshift of the excitonic transitions with decreasing barrier width, since the coupling between neighbouring wells increases and approaches asymptotically the 3D case. However, it was found that the rate of the free exciton energy downshift is similar to the bound exciton downshift, i.e. the binding energy of the acceptor bound exciton remains essentially the same all the way down to a barrier width of $L_{\mathrm{b}} \approx 10\,\text{Å}$. Furthermore, the dependence of the binding energy of the acceptor states on the barrier width has been determined, by making use of the two-hole transition satellite spectroscopy (see Sect. 6.2.7). It is found, that the $1\mathrm{S}(\Gamma_6)$–$2\mathrm{S}(\Gamma_6)$ acceptor energy does not change with the barrier width until $L_b \approx 20\,\text{Å}$, as illustrated in Fig. 6.27. Accordingly, the

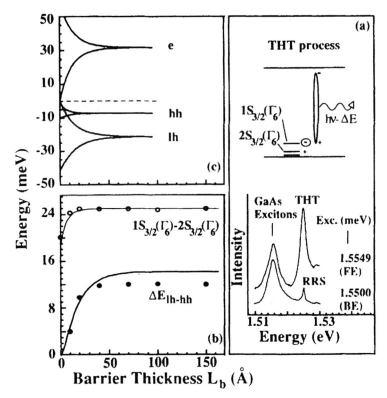

Fig. 6.27. (a) The top figure shows a schematic diagram of the two-hole-transition process. The selective photoluminescence spectra below are shown for a multiple quantum well structure with 100 Å wide wells and 40 Å wide barriers, measured at 2.0 K with the excitation energy resonant with the 1S heavy hole free exciton and the acceptor bound exciton energy, respectively. **(b)** The $1S_{3/2}(\Gamma_6)$–$2S_{3/2}(\Gamma_6)$ transition energies of the acceptors (*open circles*) and the energy separation between the 1S light hole and 1S heavy hole free excitons (*solid circles*) vs. the barrier thickness (L_b) of the structures. The *solid line* for ΔE_{lh-hh} is the calculated energy separation between the top of the first light hole and the first heavy hole minibands. The *solid line* for $1S_{3/2}(\Gamma_6)$–$2S_{3/2}(\Gamma_6)$ transition is only a guide for the eye. **(c)** The first subband energy vs. the barrier thickness calculated in the Kroning–Penney approximation. The calculated energies are given relative to the GaAs conduction band for electrons, and the GaAs valence band for holes

hydrogenic levels of the confined acceptor are not significantly affected by a neighbour well approaching. This fact implies that the bound exciton binding energy is essentially determined by the acceptor hydrogenic potential.

The dynamical properties of the acceptor bound exciton were investigated in a similar way. The decay measurements for the bound exciton was performed with the excitation resonant with the heavy hole state of the free exciton. The resulting measured decay times are presented in Fig. 6.28 (the

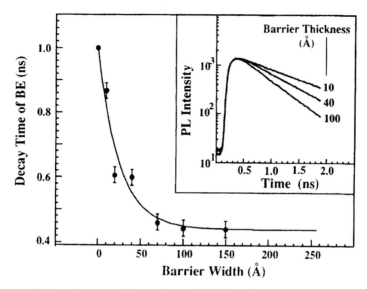

Fig. 6.28. The luminescence decay times of the bound exciton as a function of the barrier thickness, L_b, measured at 2.0 K. The inset shows three typical decay curves for some different barrier thicknesses of the acceptor bound exciton recombination for excitation resonant with 1S heavy hole free exciton

decay result for zero barrier width corresponds to the GaAs bulk value). The solid line in Fig. 6.28 is calculated by fitting the dependence of the decay time, τ, on the barrier width according to

$$\tau = \tau_\infty + (\tau_0 - \tau_\infty)\exp(-L_b/R_a) \qquad (6.26)$$

where τ, τ_∞ and τ_0 are the decay times of the bound exciton in a quantum well with a barrier width L_b, infinite barrier width (isolated quantum well) and no barrier (bulk), respectively. R_a is a fitting parameter, which provides a rough estimate of the bound exciton Bohr radius (approximately 150 Å). The bound exciton decay time increases drastically as the barrier width, L_b, decreases below ≈ 70 Å, from 400–500 ps to 1 ns for bulk GaAs (Fig. 6.28). With decreasing barrier width, the electron wave function becomes more delocalized, which in turn results in an increasing decay time. This fact gives some important insight into the electron-hole correlation of the exciton state as the two involved particles have different extension in the real space. When the barrier width decreases (below approx. 60 Å [176]), the electron states start forming a miniband, which in turn means that the electrons gains freedom in the confinement direction. The hole, on the other hand, is localized at the acceptor. Consequently, the interaction between the electron and the hole is reduced, when the barrier width decreases. This will result in a decreasing overlap between the wave function of the exciton and the acceptor wave function and accordingly a longer bound exciton decay time.

The dramatic increase of the bound exciton decay time with decreasing barrier thickness, while the binding energy remains unchanged, is an important illustration of the independence of these two parameters in the quantum well structures. It should be noted that for a quantum well both the oscillator strength and the binding energy of the bound exciton increases with increasing confinement (decreasing well width) [58]. In bulk GaAs it has been suggested that a long bound exciton lifetime (i.e. a weak oscillator strength) correlates with a large bound exciton binding energy [174]. This is obviously not completely true for the quantum wells as illustrated above. The conclusion from the above results is that the oscillator strength of the bound exciton in a quantum well is related to the relative extension of electron and hole wavefunctions, and does not necessarily correlate with the bound exciton binding energy.

Intra-acceptor Dymanics in Quantum Well Structures

The dynamics of intra-acceptor level scattering have also been studied by means of pump-and-probe measurements in AlAs/GaAs multiple-quantum well structures [177]. The experimental technique used to measure the scattering rate was a balanced pump-probe system using a free-electron laser as a source for intense far-infrared picosecond pulses. The results show that the hole relaxation time from the 2p state to the 1s ground state of the acceptor is significantly shorter, when the acceptors are confined in the quantum well structures. It was found that in bulk GaAs, the hole 2p–1s transition corresponds to a relaxation time of the order of 350 ps independent of temperature up to the thermalization temperature of the acceptor. The result from acceptors confined in the 150 Å δ-doped GaAs/AlAs quantum wells shows that the relaxation time of the excited state is of the order 80 ps.

7 The High Doping Regime

The properties of quantum well structures are strongly influenced by the presence of dopant impurities either in the well or in the barrier. As long as the impurity concentration is low, the wave functions of the impurities are spatially separated and the energy levels of the associated states are discrete. However, when the impurity concentration increases, an overlap of the impurity wave functions will take place. Initialized by the higher states a broadening of the allowed energy levels will result. This gives rise to the formation of an impurity band split off from the lowest conduction/valence band. The dependences of the impurity bands on several quantum well parameters have been calculated by J. Serre et al. [178], for the n-type case. At high doping levels, the impurity band will overlap with the free carrier continuum to form a band tail. At this point, a plasma of free carriers is formed, which can move in the potential of the fixed ions of opposite charge. This level is ususaly refered to as the metallic limit or the Mott transition [179], i.e., the electronic phase change from an insulating to a metallic behavior.

The major part of the fundamental research has been focused on modulation-doped structures, i.e. quantum wells with the impurities in the barrier, with their potential for the direct study of many-body effects, due to the high carrier concentration in combination with the maintained high mobility level [180–184]. The fascinating physical phenomena, the so-called Fermi-edge singularity has been observed in the n-type modulation doped quantum well structures. Quantum well structures doped within the well, sometimes denoted anti modulation-doped quantum wells, have important implications for quantum devices. The possible effects of impurities within the well are manifold: They can provide carriers for the conduction/valence band, act as scattering sites, hence limiting the mobility, or be important as centers for radiative or nonradiative recombination.

For low or moderate impurity concentrations, there is a sound understanding and an appropriate theoretical description of the dopants in the well. When the density of the dopants becomes higher, the situation becomes more complicated. The single particle interaction picture is not sufficient to describe such a system, since many-body interaction effects instead become increasingly more important. In case of low or moderate doping levels, the electron-hole interaction is normally limited to a two-particle effect; just the electron and hole interacting via the mutual Coulomb interaction. This fact is a consequence of the screening between the carriers. Once the impurity concentration is raised to such a level that an eqlubrium population is formed, the interaction between an individual minority carrier and the collective of the majority carriers of opposite sign, the Fermi sea, has to be taken into account. This fact results in a considerably more complex situation for the theoretical treatment. The most important deviations from the one-particle approximation are taken into account via the Hartree correction term, the Fock term (or the exchange term) and the correlation correction.

The Hartree term includes the electron-electron (or hole-hole) Coulomb interaction. The exact many-electron (hole) potential is approximated by an average potential. The Hartree contribution is usually determined by solving the Poisson equation.

The repulsion between the majority carriers of the same sign due to the Coulomb interaction, can be treated as an efficient reduction of the average charge density locally, or expressed in another way, an induced virtual charge of opposite sign, a so-called correlation particle. Accordingly, the efficient separation between the carriers will increase and the resulting rearrangements of the carriers will give rise to a lowering of the carrier energies. Due to this energy lowering, the near bandgap luminescence will be red-shifted with increasing carrier concentration. This shrinkage of the fundamental band gap as well as the higher subbands, the so-called band-gap renormalization as a result of the exchange and correlation interactions induced by the high carrier concentrations, can directly be monitored in optical spectra, as will be further expounded below.

When there are free carriers in the quantum well, the optical properties of the quantum well are modified in a significant way in various respects. This fact has important outcome for various optical devices based on such structures and it is accordingly essential to understand the associated radiative mechanisms. Modulation doped structures are significant examples on structures with an equilibrium population of free carriers. The high doping level will e.g. give rise to filling and many-body effects, but will also affect the survival of the excitons. At a sufficiently high doping level, the impurity band will overlap with the free carrier continuum. This level corresponds to the metallic limit, i.e. the electronic phase change from an insulating to a metallic behavior.

In the next, the basic concept for theoretical predictions on many-body effects will be presented. If the case of an equilibrium population of electrons is treated, the Schrödinger equation for N-electrons wave function, $\Psi(r_1, s_1, r_2, s_2, \ldots r_N, s_N)$, is employed. The *interaction between electrons* can be described by

$$H\,\Psi = \sum_{i=1}^{N} \left(-\frac{\hbar^2}{2m} \nabla^2 \Psi - Z\,e^2 \sum_{R} \frac{1}{|\,r_i - R\,|}\,\Psi \right) + \frac{1}{2} \sum_{i \neq j} \frac{e^2}{|\,r_i - r_j\,|}\,\Psi = E\,\Psi.$$

$$(7.1)$$

The second term represents the attractive potential from the nuclei at the coordinate R of the Bravais lattice and the last term depicts the mutual interaction between the electrons. If the potential from the remaining electrons is treated as a smooth distribution of negative charge with a density of ρ, the potential energy corresponding to this field is given by

$$U^{\text{el}}(r) = e \int dr_1 \rho(r_1) \frac{1}{|r - r_1|} , \qquad (7.2)$$

where the contribution from one electron to the charge density is

$$\rho_i(r) = -e \, | \, \Psi_i(r) \, |^2 . \qquad (7.3)$$

In order to derive the total electronic charge density, all electron levels should be summed up:

$$\rho(r) = \sum_i \rho_i(r) = -e \sum_i | \, \Psi_i(r) \, |^2 . \qquad (7.4)$$

With the (7.2) and (7.4), one arrives at the one-electron equation

$$-\frac{\hbar^2}{2m} \nabla^2 \Psi_i(r) - U^{\text{ion}}(r) \Psi_i(r)$$

$$+ \left[e^2 \sum_j \int dr_1 \, | \, \Psi - i(r_1) \, |^2 \, \frac{1}{|r_i - r_j|} \right] \Psi_i(r) = \epsilon_i \, \Psi_i(r) . \qquad (7.5)$$

This set of equations, usually refereed to as the Hartree equations, describes the interaction between an individual electron and the remaining electrons represented by the field as obtained by averaging the potentials of these electrons. For a more proper treatment of the electron-electron interaction, the antisymmetri required by the Pauli principle should be included in the calculations. This is done by generalizing the Hartree equations into the Hartree-Fock equations, in which an additional *exchange* term is included.

When charge carriers are exposed to an electric field, they will respond by moving into a new charge distribution. This redistribution will result in a stabilisation of the charge in such a way that there will be an overall cancellation of the electric field, when viewed from a long distance. For instance, if the eleectric field is caused by the charge distribution from an impurity, $\rho_{\text{imp}}(r)$, with a net charge of q_{imp}, this impurity will attract mobile charge carriers corresponding to a net charge of $-q_{\text{imp}}$ to its surrounding. This effect is usually referred to as the *screening*. If the charge distribution associated with the screening is given by $\rho_{\text{scr}}(r)$, the screened potential of the impurity and the screening charge is given by

$$\varphi(r) = \int \frac{\rho_{\text{imp}}(r_1) + \rho_{\text{scr}}(r_1)}{|r - r_1|} d^3 r_1 . \qquad (7.6)$$

The impurity potential is derived from

$$-\nabla^2 \, \varphi_{imp}(r) = 4\,\pi\rho_{imp}(r) \tag{7.7}$$

and is assumed to be linearly related to the screened potential according to

$$\varphi_{imp}(r) = \int \epsilon\,(r,r_1)\,\varphi(r_1)\,dr_1\,, \tag{7.8}$$

where ϵ is the dielectric constant of the material. Then, in order to determine the charge density in the presence of the screened potential, the electronic density should be constructed from the one-electron wave functions via (7.4) and after that the one-electron Schrödinger equation

$$-\frac{\hbar^2}{2m}\,\nabla^2\,\Psi_i(r) - e\,\varphi(r)\,\Psi_i(r) = \epsilon_i\,\Psi_i(r) \tag{7.9}$$

should be solved by a linear order of perturbation theory.

For the case of p-doped bulk GaAs, B. Sernelius [185,186], has calculated the bandgap renormalization in the high doping regime including exchange and correlation effects of the electron-hole system together with the interaction with ionised impurities.

The hole subband dispersions can be determined by a self-consistent calculation of the Schrödinger and Poisson equations. The kinetic energy operator is represented by the Luttinger–Kohn Hamiltonian [107,187], which includes the interaction between the heavy and light hole.

$$H^{kin} = \begin{pmatrix} P+Q & L & M & 0 \\ L^+ & P-Q & 0 & M \\ M^+ & 0 & P-Q & -L \\ 0 & M^+ & -L^+ & P+Q \end{pmatrix}, \tag{7.10}$$

where

$$P = \frac{\gamma_1\hbar^2}{2m_0}\,k^2\,,$$

$$Q = \frac{\gamma_2\hbar^2}{2m_0}\,(k_x^2 + k_y^2 - 2k_z^2)\,,$$

$$L = -i\frac{\sqrt{3}\,\gamma_3\hbar^2}{2m_0}\,(k_x - ik_y)\,k_z\,,$$

$$M = \frac{\sqrt{3}\,\hbar^2}{4m_0}\,(\gamma_2 + \gamma_3)\,(k_x - ik_y)^2\,.$$

The γ_1, γ_2 and γ_3 are the Luttinger parameters describing the Γ_8 valence band. In this development, the axial approximation [188] with an average dispersion in the xy-plane, has been employed. This corresponds to replacing

γ_2 and γ_3 in term M in the matrix above by $(\gamma_2$ and $\gamma_3)/2$. Both matrices give the same subband structure corresponding to a two-fold spin degeneracy for the case of symmetric potential. However, a full many body theory for 2D holes is complex and some approximations have to be adopted. For instance, the Hartree approximation is applied in the subband calculations. Non-parabolicities in the dispersion of the bands have been neglected. The exchange and correlation effects are included as described below. Further, only the uppermost $n = 1$ heavy hole subband is assumed to have an equilibrium population of holes, i.e. the Fermi level is situated between the heavy hole and light hole subbands. Due to momentum conservation, an absorption process between the $n = 1$ heavy hole and the $n = 1$ electron subbands can just take place at the Fermi wave vector, $k = k_F$.

The three different types of carriers, i.e. electrons, heavy holes and light holes are assumed to move in a gas of heavy holes. Due to the Fermion statistics and the repulsion between each pair of heavy holes and a heavy hole is surrounded by an exchange-correlation hole. This exchange-correlation hole is negatively charged, while the heavy hole is positively charged. This interaction between the charged species results in a negative energy contribution, a self-energy shift.

If we instead focus on the light hole, this hole is assumed to be surrounded by a correlation hole due to the repulsion between light holes and heavy holes. In this case, there is no exchange contribution. The interaction between the correlation-hole and the light hole also results in a negative energy shift.

For the electron, finally, an enhancement of the heavy hole concentration due to the attractive force between the electron and heavy holes is assumed to surround the electron, which in turn results in a negative energy shift. Accordingly, we can conclude that all three types of particles attain negative energy shifts due to the interaction with the heavy hole gas.

Similar energy shifts are obtained from the interaction with the acceptors. While both heavy holes and light holes are repelled by the acceptor ions and tend to avoid them, the electrons are attracted by the acceptor ions. This leads to a reduction of the energy for all types of carriers. The many body effects are obtained in perturbation theory on the states determined beforehand, self-consistently neglecting the many body effects.

7.1 Band Filling Effects

When the doping is sufficiently high, the Fermi level will pass the continuum edge, which in turn will result in an equilibrium population of carriers, as mentioned earlier. Accordingly, the first available state in an absorption process will be up-shifted to the Fermi level at lowest temperature. In an allowed optical band-to-band absorption process, both the initial state in the valence band and the final state in the conduction band must have the same k-vector, k_F in order to conserve momentum. Accordingly, this will result in

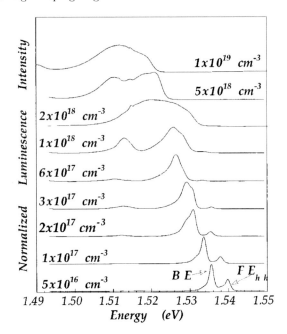

Fig. 7.1. Photoluminescence spectra for acceptor center doped 150 Å wide GaAs/Al$_{0.3}$Ga$_{0.7}$As quantum wells with different doping concentrations. All spectra are normalized. The acceptor bound excitons dominate even at low doping levels. The heavy hole free exciton intensity decreases continuously with increasing doping, while the bound exciton gains intensity

an increasing blue-shift of the absorption process with increasing doping. In emission, on the other hand, there is a red-shift appearing due to the bandgap renormalization (Fig. 7.1). The emission process occurs at $k = 0$ and consequently there will be an increasing energy gap between the absorption and emission transitions, the Burstein–Moss shift, with increasing doping.

The bandgap renormalization is found to dominate relatively band-filling effects in the p-type quantum wells, which results in the photoluminescence red shift of the emission with increasing doping. Also the edge observed in photoluminescence excitation spectrum is redshifted. However, the experimentally observed energy separation between the light hole and heavy hole bands, or more correctly the excitons associated with these subbands, monitored in a photoluminescence excitation spectrum, decreases as the hole concentration increases [189, 190]. The evolution of the photoluminescence excitation spectra with increasing hole concentration is shown in Fig. 7.2 for a 150 Å wide quantum well, which is doped with acceptors in the center of the well. It is obvious that the observed energy separation between the excitons associated with the light hole and heavy hole bands, respectively, decreases significantly with increasing doping level might, e.g., from 1.544 eV

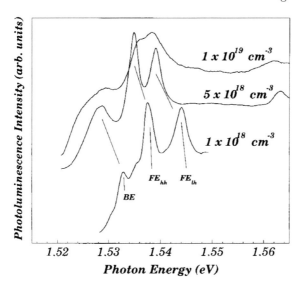

Fig. 7.2. Photoluminescence excitation spectra for acceptor center doped 150 Å wide GaAs/Al$_{0.3}$Ga$_{0.7}$As quantum wells for some different doping levels, measured at 2 K. The excitons survive up to the highest doping concentration shown, as clearly seen in these spectra

to 1.534 eV, as the doping level is increased from 10^{14} up to 10^{18} cm^{-3}. This striking effect is mainly due the band-filling effect, which has to be taken into account in this case [189, 190].

Due to the conservation of momentum, an excitation process involving a hole state at the Fermi level, i.e., with a momentum of $k = k_F$, has to involve an electron also of a momentum of $k = k_F$. A comparison with the theoretical predictions including both bandgap renormalization and band-filling, as shown in Fig. 7.3 exhibits an excellent agreement with the experimental results all the way up to a doping level of approximately 10^{17} cm^{-3}.

At the very highest doping levels measured, there are still some deviations between the theoretical predictions and experimental results. In order to make a proper theoretical prediction also at these hole concentrations, a more advanced model should be employed. In such a more sophisticated model, one can not allow any significant approximations, which are not adequate in the high doping regime. For instance, one has to include effects of filling also the light hole. The 2D-approximation becomes less good (just relevant for doping levels up to about 2×10^{18} cm^{-3}). The neglect of the interband transitions is a very good approximation for transitions across the band gap, due to the large energy difference in the denominator in the contributions from these transitions. For transitions between the hole-bands this approximation may be questionable, especially for higher doping levels, when the Fermi level approaches the top of the light hole band.

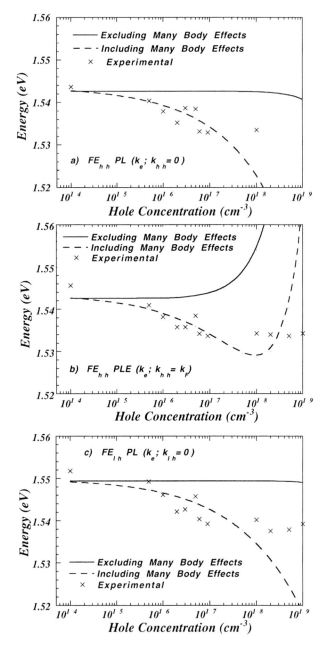

Fig. 7.3. The calculated dependencies of the peak positions for the free exciton on the hole concentration for (**a**) the heavy hole free exciton (FE$_{hh}$) in luminescence, (**b**) the FE$_{hh}$ in photoluminescence excitation spectrum, (**c**) the FE$_{lh}$ in photoluminescence excitation spectrum. The theoretical predictions are furthermore compared with the experimental results derived from the luminescence measurements

Also the dependence of the exciton binding energy on the doping level has to be taken into account, since the fundamental electron-hole interaction, the exciton, is also affected at higher carrier concentrations. This effect is usually referred to as exciton screening. The excitons can be screened by either other excitons or by free carriers. Due to the onset of this screening and the exclusion principle, the fundamental excitons become less stable at moderate densities with a decreasing exciton binding energy as a result and eventually unbinding of the exciton state [191]. The most important factors affecting the screening effect are the *phase space filling, exchange and Coulomb screening*. When a state in the *phase space* is occupied by a free carrier, this state is excluded from formation of excitons, which in turn results in a reduction of the exciton binding energy. Furthermore, the carriers are not uniformly distributed, but due to the inter-particle Coulomb repulsion, the carriers become spatially correlated, which results in a lowering of the energy. At high carrier concentration instead of the fundamental exciton, another exciton state originating from the interaction between many carriers in the Fermi sea and a photogenerated carrier of opposite charge gives rise to the so called Fermi edge singularity, as will be further expounded in the following section.

The screening of 3D excitons is strong already at low doping concentrations and carrier densities (say $10^{16}\ cm^{-3}$), while the corresponding screening effect in the 2D case is much weaker. The restriction on the movement of carriers due to the lower dimensionality inhibits their ability to screen. For the case of quantum wells, excitons are expected to survive all the way up to the degenerate limit [192,193]. Consequently, a major difference between the 2D and 3D system is the relatively stronger exciton recombination in optical spectra for the 2D case even at high carrier concentrations.

At even higher concentrations, the excitons finally become completely quenched. The transmission measurements in a range of undoped through heavily modulation-doped GaAs/AlGaAs multiple quantum well structures have been performed by D. Huang et al. [194]. They demonstrated experimentally the quenching of the excitonic oscillations with an increasing density of the quasi-two-dimensional electron gas. The electron density corresponding to the total bleaching of the lowest excitonic transition is greater than or equal to $3 \times 10^{11}\ cm^{-2}$ for a quantum well size of $200\ \text{Å}$. Theoretical calculations of the absorption spectra which include the effect of carrier screening have also been made [195]. The results show that both long-range and short-range many body effects should be included to explain the experimentally observed spectra. In the modulation-doped case, they conclude that the phase-space filling and exchange of the electron gas are the dominant effects on excitonic absorption. The results show that the exciton lifetime reduces due to the interaction between the electrons and the excitons.

7.2 Bandgap Renormalization

At a sufficiently high doping concentration, the impurity band will overlap with the free carrier continuum. This level corresponds to the metallic limit, i.e. an electronic phase transition from a semiconducting to a metallic behaviour. The exchange correlation interactions, induced by the high carrier concentrations, give rise to a decrease of the fundamental band-gap, i.e., the bandgap renormalization. With increasing carrier concentration, there are two major counteracting effects affecting the absorption transition above the band-gap: A blue shift due to the band filling up to the Fermi level, the "Burstein–Moss" shift, and on the other hand, a red shift caused by the bandgap renormalization. These band-gap shifts (of the fundamental absorption edge) depend on, e.g., the semiconductor material, the type of doping and the carrier concentration. The Burstein-Moss shift and the shift caused by bandgap renormalization are of different sign, but of comparable magnitude. For n-type GaAs, the Fermi level shift dominates [186, 193, 196–198], while the reverse situation applies to p-type GaAs [197, 199–202].

The hole subband dispersion can be determined by a self-consistent calculation of the Schrödinger and Poisson equations. The kinetic operator is represented by the Luttinger–Kohn Hamiltonian [107, 187], which includes the interaction between the heavy and the light holes. This is generally a 4×4 matrix but can be decoupled into two 2×2 matrixes [203]

$$\begin{pmatrix} A_+ & C \mp iB \\ C^* \pm iB & A_- \end{pmatrix} , \qquad (7.11)$$

where

$$A_\pm = -\frac{\hbar^2}{2m_0} \left[(\gamma_1 \pm \gamma_2) \left(k_x^2 + k_y^2 \right) + (\gamma_1 \mp 2\gamma_2) \, k_z^2 \right] ,$$

$$B = \frac{\hbar^2}{2m_0} \left[2\sqrt{3}\, \gamma_3 \left(k_x \, k_z - i \, k_y \, k_z \right) \right] ,$$

$$C = \frac{\hbar^2}{2m_0} \left[-\sqrt{3}\, \gamma_2 \left(k_z^2 - k_y^2 \right) - 2\sqrt{3}\, \gamma_3 \, i \, k_x \, k_y \right] .$$

Here the axial approximation [188] with an average dispersion in the xy-plane is employed. Instead of the γ_2 and γ_3 parameters in terma B and C above, the term $(\gamma_2$ and $\gamma_3)/2$ is adopted. Both matrices give the same subband structure corresponding to the twofold spin degeneracy for a symmetric potential. Furthermore, effects of non-parabolicity should be taken into account (see, e.g., [204]) and relevant boundary conditions (see, e.g., [205]) should be fulfilled. There are three different kinds of particles present: Electrons, heavy holes and light holes, characterized by the effective masses, m_e^*, m_{hh}^* and m_{lh}^*, respectively, which are assumed to move in a gas of heavy holes. Due to the prevailing Fermion statistics and the repulsion between each pair of heavy holes, the positively charged heavy hole is surrounded by an effective negative charge, which will be denoted an exchange-correlation hole. Accordingly, the interaction between these charged species leads to a negative energy contribution – a negative self-energy shift. For the light hole, there is no exchange but a correlation contribution. This correlation effect is due to the repulsion between light and heavy holes. The resulting interaction between the correlation-hole and the light hole leads also in this case to a negative energy shift. Finally, the electron is surrounded by an enhanced heavy hole concentration caused by the attraction between electrons and heavy holes. Again, the interaction gives rise to a negative energy shift. Accordingly, all three types of particles attain negative energy shifts due to the interaction with the gas of heavy holes.

The effect on the energy shift due to the interaction with the acceptors is similar. Both the light and heavy holes are attracted by the acceptor ions and stay close to them in contrast to the electrons, which are repelled by the ions. Consequently, the interaction with the acceptors results also in this case in a reduction of the energy for all three types of particles. The essential difference between the particle interactions with the heavy holes in comparison with the ions is that the latter are fixed in position.

The many-body shifts can be derived in a 2D approximation, performed in a pure Random Phase Approximation with Hubbard's local field correction [189, 190]. It turns out that the inclusion of a local field correction has only minor effects on the results. The results we present here are those with the local field correction included. The many-body effects are derived within the Rayleigh–Schrödinger perturbation theory, or on-the-mass-shell perturbation theory [202]. In this theory, the single-particle energy for a state $(\boldsymbol{k}, i)$ is defined as the variational derivative of the total energy with respect to the occupation number for state $(\boldsymbol{k}, i)$, i.e.,

$$\epsilon_{\boldsymbol{k},i} = \frac{\delta(E_{\text{tot}})}{\delta n_{\boldsymbol{k}}^i} = \epsilon_{\boldsymbol{k},i}^0 + \hbar \sum_{\boldsymbol{k},i}, \tag{7.12}$$

where E_{tot} is the total energy. The first term on the right hand side is the unperturbed single-particle energy (the kinetic energy) and the second, the self-energy from the interactions. The index i runs over the electrons, heavy and light holes. The total interaction energy can be written as

$$E_{\text{xc}} = +i\frac{1}{2}\int_0^1 \frac{d\lambda}{\lambda} \sum_{\boldsymbol{q}} \left\{ \int_{-\infty}^{\infty} \frac{d\omega}{2\pi} \hbar \left[\frac{1}{\epsilon^\lambda(\boldsymbol{q},\omega)} - 1 \right] - \frac{(N_e + N_{\text{hh}} + N_{\text{lh}})\lambda \nu_q}{i\nu\kappa} \right\}, \tag{7.13}$$

where $\nu_q = \frac{2\pi e^2}{q}$ is the 2D Fourier transform of the Coulomb potential, λ is the coupling constant for the interaction, ϵ is dielectric function is the test-particle function, i.e.,

$$\epsilon^\lambda(\boldsymbol{q},\omega) = 1 + \lambda \alpha_{0,e}(\boldsymbol{q},\omega) + \lambda \alpha_{0,\text{lh}}(\boldsymbol{q},\omega) + \frac{\lambda \alpha_{0,\text{hh}}(\boldsymbol{q},\omega)}{1 - f_{\text{H}}(\boldsymbol{q})\lambda \alpha_{0,\text{hh}}(\boldsymbol{q},\omega)}. \tag{7.14}$$

Here α is the polarizability, for the light holes and electrons, respectively. The variational derivative with respect to the occupation numbers for these particles will produce the corresponding self-energy shifts. When the derivative has been taken, these polarizabilities can be put equal to zero. The polarizabilities originate from the 2D Random Phase Approximation (RPA) functions. When the 2D form of the Hubbard's local field correction:

$$f_{\mathrm{H}}(\boldsymbol{q}) = \frac{1}{2}\sqrt{\frac{q^2}{q^2 + k_{\mathrm{F,hh}}^2}} \tag{7.15}$$

is equal to zero, a pure Random Phase Approximation is obtained when this function is put equal to zero.

The resulting band gap renormalization as a function of hole concentration for a 150 Å wide quantum well is illustrated in Fig. 7.4. The dependencies of the electron, light and heavy hole subbands shift with the hole concentration at $k = 0$ together with the corresponding dependencies of the electron and heavy hole subbands at the Fermi wave vector (k_{F}) are shown in the figure. From this figure it is obvious that the band gap renormalization is very sensitive to the hole concentration. The heavy hole band exhibits a stronger bandgap renormalization effect than the light hole band. When comparing with optical spectra, e.g., in absorption and emission spectra, one must keep in mind that the optical transitions involve also electrons. In order to keep momentum conservation, electrons with the same k vector as the hole takes part in an allowed optical transition. The subband shift from the bottom of the band ($k = 0$) out to the Fermi level ($k = k_{\mathrm{F}}$), the phase space filling, is significantly larger for the electron subband than for the heavy hole subband. Consequently, this effect would contribute to an increasing energy shift between the transitions monitored in photoluminescence and in photoluminescence excitation spectra with increasing hole concentration.

7.3 Many Body Effects

When solving the Poisson equation, the Fermion character of the holes and electrons has been neglected. The inclusion of the Fermion character results in the exchange correction. Furthermore, due to the Coulomb interaction, the carriers will be spatially correlated. Due to this process, generally referred to as the correlation interaction, the carrier energy can be lowered.

The effects caused by many body interactions can be incorporated in different ways. The simplest way to include many-body effects in the calculations is to add an exchange-correlation potential to the Hamiltonian in the self-consistent calculation of the Schrödinger and Poisson equations. This potential is dependent on the 3D density of holes and is the same for the different sublevels. The result of the addition of this exchange-correlation potential, is a minor change in the dispersion of the bands. In such a treatment,

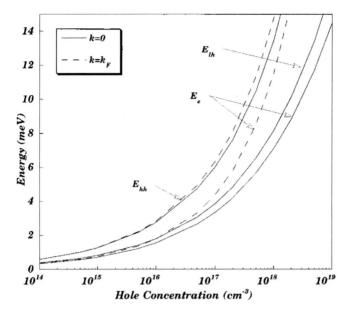

Fig. 7.4. The theoretically predicted renormalization of the electron, heavy-hole, and light-hole subbands with the hole concentration at the bottom of the band ($k = 0$) together with the corresponding dependencies of the electron and heavy-hole subbands at the Fermi wave vector (k_F) are shown in the figure

all states in a band will experience the same many-body potential and are energy shifted in a similar way.

An alternative way to include the many body interactions is to take the Hartree contribution and the contribution from the carrier-carrier scattering into account on a similar basis. Then the energy shifts can be calculated by perturbation theory [205]. A third way is to start to solve the Hartree problem and then use the so-derived solution as a basis for a perturbation treatment including the carrier-carrier scattering. The important contribution for the many body interaction originates from the scattering processes involving small momentum transfer (typically smaller than twice the Fermi momentum). This fact defines an upper limit for the hole concentration what concerns the validity of a strict 2D approach.

If we furthermore neglect interband transitions, the involved particles can be treated as three different particles: Electrons, heavy holes and light holes in the self-energy calculations. These particles are then assumed to move in a heavy hole gas. Furthermore, each heavy hole can be supposed to be surrounded by an exchange-correlation hole due to the Fermion statistics applicable and the repulsion between the holes. This exchange-correlation hole is negatively charged. The interaction with the positively charged heavy hole results in a *negative* energy contribution for the heavy hole, a self-energy shift. A light hole can be presumed to be surrounded by a correlation hole due to the repulsive interaction between the heavy holes and the light holes.

Another approach is to employ the self-consistent Hartree results including non-parabolicity caused by valence-band mixing as a starting point for the many-body calculations. Furthermore, in order to make the calculations, one often performs them in a quasi-two-dimensional approximation, i.e., neglecting the finite extension of the wave functions perpendicular to the well. This calculation is made to find out the effects from having a finite well width of the many-body self-energy shifts for the different states: In this approach states at the Fermi-level, at the valence-band maximum and at the bottom of the conduction band are all shifted with different amounts.

The result of the theoretical treatment on the many-body effects in order to predict the energy position of the free excitons is plotted as a function of the hole concentration in Fig. 7.3. As can be seen in this figure, the heavy and light hole related free excitons have a similar dependence on the hole concentration. In Fig. 7.3, the theoretically predicted dependencies of the transition energies involving heavy holes and light holes, respectively, on the hole concentration are compared with experimental results as derived from photoluminescence and photoluminescence excitation measurements. It is obvious from this figure that many-body effects become important already at low hole concentrations (below 10^{16} holes per cm^{-3}). Also the result of the theoretical model employed appears to be in nice agreement with the experimental results up to high hole concentrations ($\approx 2 \times 10^{18}$ holes per cm^{-3}), despite the significant approximations adopted, which are not adequate in the high hole concentration regime. For instance, the application of a true 2D approximation and a constant exciton binding energy do not hold for hole concentrations exceeding 2×10^{18} cm^{-3}. Another fact to be taken into account at high impurity levels is the impurity band mergence with the subband to form a band tail [178]. Also, for high densities, the light hole band will be filled (for densities above $\approx 3 \times 10^{18}$ cm^{-3}) and the actual dispersion curve for the heavy hole band becomes flatter for higher momentum. The many-body effects become increasingly important with increasing effective mass and the neglect of non-parabolicity effects accordingly underestimates the many-body effects for high hole concentrations.

An example on the effect bandgap renormalization for the case of a 150 Å wide quantum well in the high doping regime according to theoretical predictions as described above is shown in Fig. 7.4. These theoretical predictions will next be compared with experimental results as provided from photoluminescence measurements at low temperatures of highly p-type doped quantum wells. Figure 7.1 displays the development of luminescence spectra with increasing hole concentration. The photoluminescence spectra are dominated by two excitonic features, the heavy hole related free exciton (FE^{hh}) and the neutral acceptor bound exciton. This series of luminescence spectra clearly demonstrates, that the exciton peaks are redshifted as the hole concentration increases. If these experimental results on the free exciton redshift are compared with the theoretical predictions for the same quantum well structure (Fig. 7.3), one can conclude that a close agreement with the experimental results is achieved, except for the very highest doping levels measured.

7.4 Fermi Edge Singularity

This many electrons-one hole interaction for n-type material (or many holes-one electron for p-type material) results in the formation of the so-called Mahan exciton or the Fermi-edge singularity [180–182]. This many particle interaction results in a minimization of the hole (electron) energy and an associated enhancement of the oscillator strength for optical transitions close to the Fermi-edge [182–184].

When a semiconductor is heavily doped to concentrations beyond the metallic limit, interesting many-body phenomena appear. One of the best known application examples on such a structure is the modulation doped quantum well or heterostructure. In this case, a charge transfer to the well or notch potential, respectively, will occur until an equilibrium situation is achieved, when the carriers have filled up to the Fermi energy. Accordingly, the free carriers are spatially separated from the parent impurity. Due to this spatial separation, such structures exhibit a minimum impurity scattering and consequently a high mobility. This important property is utilized in several applications, such as High Electron Mobility Transistors (HEMTs). From magneto-transport experiments, such as Hall or Schubnikov de Haas measurements, important parameters like the carrier concentration and mobility can be evaluated. Also fundamental physical aspects, like the quantization of the Hall effect, can be reviewed in magneto-transport measurements. However, the properties of such a semiconductor can preferably be studied by optical means as well. Upon optical excitation, additional electron-hole pairs are generated. However, the increase of the majority carrier equilibrium population due to the optical pumping is negligible, while the minority carrier population is determined by the optical excitation.

7.5 Charged Excitons

The quantum well related luminescence is usually dominated by intrinsic procsses. Due to the electron-hole Coulomb interaction, bound states, excitons are formed. These excitons are mobile in the lattice and are usually denoted free excitons. The free excitons are predominant in optical spectra measured at lowest temperatures, while band-to-band transitions gain intensity with increasing temperatures, in intentionally undoped samples as mentioned above. The *free exciton* processes are well documented in the literature. The existence of stable three particle states, when there is an excess carrier population of electrons or holes, was also predicted by M.A. Lampert already 1958 [62]. This excitonic interaction in the presence of a degenerate electron gas was first considered by G.D. Mahan [180]. The three particle state is the semiconductor complex analogous to the charged hydrogen atom, either the negatively charged H^- or the positively charged H_2^+. However, the experimental verification [206, 207] of the negatively or positively charged exciton has been weak or absent in bulk semiconductors due to the small binding energy of the third particle [208], about $0.03\,Ry^*$ (corresponding to $0.12\,meV$ in GaAs). This binding energy corresponds to the energy down shift of the charged exciton relatively the free exciton, or explained in a different way, the binding energy of a free excess electron (or hole) to a neutral exciton. This means that any weak feature originating from a charged exciton should be overlapping with and hidden by predominant exciton peaks in luminescence measurements. Since this small binding energy is associated with the extended wave function of this third particle, the confinement effect prevailing in a quantum well will give rise to a significant enhancement of the binding energy of the negatively or positively charged exciton for a 2 dimensional structure. This enhanced binding energy has been theoretically predicted [209] as well as experimentally documented both in II–VI [210] and III-V materials [211]. For instance, the binding energy of the negatively charged exciton in a $200\,\mathring{A}$ wide GaAs/AlAs quantum well is found to be $1.3\,meV$ [212]. A quantum well has in addition the advantage what concerns observation of the charged exciton that the electron (or hole) density can be controlled via modulation doping and an applied external field. The charged exciton is possible to observe at low or moderate concentrations of excess electrons (or holes), while at higher densities, the Coulomb attraction is screened [213] and the charged excitons become quenched [59].

7.5.1 Negatively Charged Excitons

In analogy with the negatively charged hydrogen atom, H^-, negatively charged excitons, X^-, can form, when there is an excess population of electrons. This excess electron density can be controlled by e.g. applying a gate voltage across an n-type modulation doped quantum well. As long as the extra charge is nearly deleted in the well, the luminescence spectrum is dominated by the normal neutral exciton, X. When the excess electron density increases, the intensity of the neutral exciton, X, attenuates, while the negatively charged exciton state, X^-, strengthens correspondingly. The X^- intensity gain is associated with the increased area of the quantum well, which is covered with excess electrons [214, 215]. Usually, the X^- recombination is observable in remotely doped quantum wells, if the excess electron concentration is of the order a few $10^{10}\,cm^{-2}$ [216]. The recombination process has the negatively charged exciton state, X^-, as the initial state and a photon together with the excess electron as the final state. This recombination will occur at an energy slightly below the neutral exciton by an energy displacement, which corresponds to the binding energy of the second electron [210, 214–216]. As the electron concentration is even higher, the neutral exciton, X, becomes completely quenched. The negatively charged exciton, X$^-$, is red-shifted due to partly the exchange-correlation potential caused by the electron-electron interaction and the electric field due to the charge in the quantum well. For higher electron densities, the X$^-$ evolves smoothly into the Fermi edge transition [210, 215, 216].

The X$^-$ is found to be very sensitive to an applied electric field due to the repulsion between the two electrons in the X$^-$ [216]. As the electric field increases, the electron and hole wave functions are polarized in opposite directions. In the photoluminescence spectrum, both the neutral exciton, X, and the X$^-$ exciton are redshifted considerably with an increasing electric field due to the Stark effect, but with a decreasing energy separation between these excitons. This means that the mutual Coulomb attraction is reduced and the X$^-$ binding energy decreases. The neutral exciton, X, on the other hand, is not affected by the electric field in such a drastic way, since the repulsive force between the electrons is missing in this case.

The decay time of the negatively charged X$^-$ exciton is significantly shorter, by a factor of 4, than for the neutral exciton at low temperatures [213], as can be expected for such a many particle transition. This estimate can not be evaluated directly from transient photoluminescence measurements, since the neutral and negatively charged excitons are in thermal equilibrium with each other and the measured decay time of the X$^-$ exciton implicates also the feeding process from the neutral exciton, X, at a given temperature. In order to evaluate the true decay time, a fitting procedure involving also the integrated photoluminescence intensity has to be employed. As the temperature increases, the decay time of the X as well as the X$^-$ exciton increases. This has been explained earlier for the neutral exciton as due to the spread of the exciton momenta as the temperature increases. Only excitons with momenta similar to the photon momentum are able to decay due to the momentum conservation restriction. This means that there is a diminishing fraction of excitons fulfilling this criterium [165].

Another consequence of the multi-particle character of the charged excitons, X$^-$, is their potentiality for shake-up processes. When an electron-hole pair recombines, there is a non-negligible probability that the second electron is ejected into a higher energy state, giving rise to a low energy satellite peak. This has been observed for the X$^-$ in the presence of a magnetic field, when the second electron is excited into a higher Landau level [217]. The resulting satellites is accordingly down-shifted relatively the X$^-$ with an energy corresponding to the energy separation to the Landau levels. Another experimental result from a shake-up process is the up-conversion, i.e. an emission observed at a higher energy than the laser excitation energy [218]. The up-converted emission observed at an applied magnetic field is related to the recombination of the second electron, which has been excited into a higher Landau level in an Auger-like process.

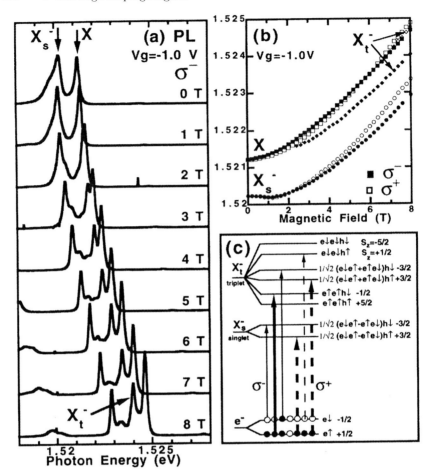

Fig. 7.5. (a) Photoluminescence spectra recorded with different applied magnetic fields, emitted for σ^--polarization, at 2.0 K and with a Schottky bias of -1.0 V. **(b)** Evolution of the luminescence peak energies (*vertical axis*) with the field, for both σ^- (*solid*) and σ^+ (*open symbols*) polarizations. Note the emergence of the triplet X_t^-, (*diamonds*) near 2.4 T in σ^--polarization in (a) and (b). **(c)** Schematic illustration of the X^- transitions, based on the absorption of circularly polarized photons in a magnetic field, $e^- + $ photon $\longrightarrow X^-$. The photon changes the total z component spin, S_z by $+1$ (-1) for σ^+ (σ^-). (From [215])

The behavior of the negatively charged exciton, X^-, in the presence of a magnetic field differs significantly from the neutral exciton, X. First of all the increasing blue-shift with increasing magnetic field, the diamagnetic shift, for the neutral exciton, X, is replaced by an initial red-shift with field for the X^-, before increasing in energy, but at a slower rate than for the X. This means that there is an increasing energy separation between X and X^-, i.e. an increasing binding energy of the second electron (by 50% up to 8 T [215]). This can be understood in terms of the X^- wave function consisting of a combination of two one-electron orbitals with different spatial extensions. The relatively weak binding of the second electron results in a significantly larger orbital for this electron. The magnetic field will affect the confinement of the spatially extensive wave function associated with the outer orbital. The confinement of this orbital will enhance the overlap and accordingly the Coulomb interaction with the core exciton. As a result, the binding energy of the second electron and consequently also of the X^- will increase with the field.

The Pauli exclusion principle requires the spin wave function of the two electron systems to be either anti-symmetric (singlet) or symmetric (triplet). In analogy with the negative hydrogen ion, the singlet state is at lowest energy. In zero-field, it is only this singlet state, which is expected to bind, while the triplet state will just bind in the presence of a moderate magnetic field (a few T) [215] as demonstrated in the polarized photoluminescence results shown in Fig. 7.5.

In the presence of a magnetic field, the lower singlet state of the X^- splits into two components, while the upper X^- triplet state also splits into two components [215]. The upper X^- triplet state is overlapping with the free exciton and is resolved from the free exciton only in the presence of a magnetic field. The observed splitting between the two X^- singlet components exceeds the energy separation of the free exciton components with a factor of $\approx$ 3 [215]. Furthermore, the reversal of sign in the Zeeman splitting seen for X around 6 T due to the mixing of the heavy- and light-hole subbands [219,220] is not observed for the X^-.

7.5.2 Positively Charged Exciton

Similarly to the negatively charged exciton, X^-, the corresponding *positively charged exciton*, X^+, has been observed in quantum wells remotely doped with acceptors [221]. Even the small background acceptor concentration in the AlGaAs barrier is sufficient to create a significant population of holes, enough to exhibit the X^+ exciton in quantum wells [222]. The X^+ exciton appears approximately 1 meV below the neutral exciton, X (Fig. 7.6), but dependent on the well width (from 0.97 meV in a 300 Å wide well to 1.35 meV in a 140 Å well [222]). A notable dependence on the excitation energy has also been observed. Due to the electric field created by the acceptor modulation doping, the photoexcited electrons and holes captured in the well behave in a different way, when the excitation energy is above or below the band gap energy of the AlGaAs barrier layer. With excitation energy lower than the bandgap of the AlGaAs barrier layer, the experimental results showed that the dependence of the X^+ on the excitation power is weak, and the X^+ is hardly quenched by just increasing the excitation density. However, when the excitation energy is above the bandgap energy of the AlGaAs barrier layer, the dependence of the X^+ on the excitation density is stronger in compari-

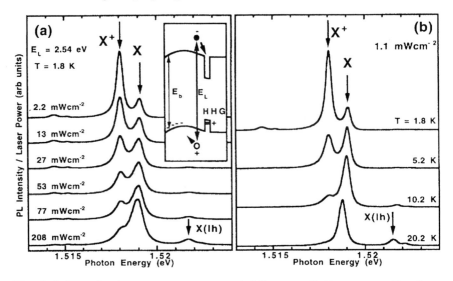

Fig. 7.6. Photoluminescence spectra recorded for an excitation energy, E_L, above the band gap energy of the barrier layer, E_b, for (a) different laser power densities and (b) different sample temperatures. The inset illustrates how the electrons, photoexcited in the upper barrier layer, are captured into the quantum well, thereby lowering the excess hole density. This explains the strengthening of the X line relative to the X^+ line with increasing laser power. (From [222])

son with the case of below bandgap excitation, and the X^+ can be quenched with increasing the excitation power in some cases [222]. The observed different behaviors are due the difference of the carrier capture. At the above bandgap excitation, the photoexcited electrons are captured into the well to recombine with the hole gas and accordingly reducing the hole density, while the photoexcited holes drift in the opposite direction, away from the well, due to the electric field and trapped at the potential trap at the maximum valence band. Therefore with increasing excitation power, the reduction of the hole density becomes stronger, resulting in a stronger dependence of the X^+ intensity on the excitation density. While the excitation energy is tuned below the bandgap of the AlGaAs barrier, on the other hand, the excess hole density will not significantly change since both electrons and holes are created in the well layer. Due to the small binding energy of the X^+ exciton, the energy separation relatively the neutral exciton, X, the X^+ exciton is thermally quenched already at temperatures of $\approx 10\,\mathrm{K}$, caused by the thermal dissociation, $X^+ \longrightarrow X + h$.

In the presence of an applied magnetic field, the X^+ exciton exhibits a blue-shift, which is approximately quadratic with the field (in the low field domaine), as characteristic for excitons. However, the X^+ exciton is blue-shifted less rapidly than the neutral exciton X and the splitting between X^+ and X increases with field (Fig. 7.7). This means that the binding energy of

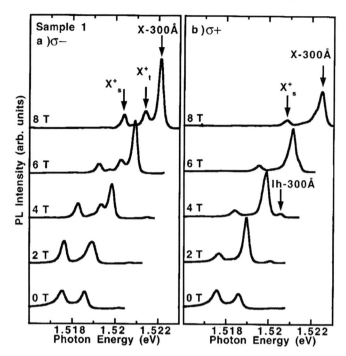

Fig. 7.7. Photoluminescence spectra emitted in (a) σ^- and (b) σ^- circular light polarization, from a 300 Å wide quantum well structure, with different magnetic fields applied perpendicular to the layers. (From [222])

the X^+ exciton increases, as the second hole binding energy of X^+ increases due to the increasing magnetic confinement. Due to the more rapid blue-shift of the neutral exciton X, another well resolved peak emerges on the low energy side of the X line at higher fields. This novel peak is due to the excited spin triplet state of the X^+ exciton, with a symmetric spin wave functions of the two holes involved [222]. The low energy component of the X^+ exciton, which is observed already at zero field corresponds then to the spin singlet state.

7.6 Effect of an Applied Magnetic Field

Magneto-luminescence is a powerful method for investigations of cardinal physical parameters, such as the effective mass, the effective Landé factor and the exciton binding energy. At small or moderate magnetic fields, the electron-hole recombination can be described as magnetoexcitons as long as the mutual Coulomb interaction exceeds the magnetic field energy. At higher fields, the magnetic field energy term becomes comparable with the Coulomb energy, i.e.,

$$\frac{\gamma_{\mathrm{B}}}{\hbar\omega_{\mathrm{c}}}(2R_0) \approx 1 \,, \tag{7.16}$$

where γ_B is the reduced magnetic field, ω_c is the cyclotron frequency and R_0 is the effective Rydberg energy. Then the electron-hole recombination will be dominated by transitions between free electron and hole Landau levels [223].

When a magnetic field is applied along an arbitrary direction in a quantum well system, the magnetic field will result in a coupling of the motions in the different directions. In this case, there is no general solution to the Schrödinger equation, but in the limits of very small or very large magnetic field, either the field or the confinement can be treated as a perturbation, which in turn means that one can derive a solution to the Schrödinger equation. However, in the presence of a magnetic field perpendicular to the plane of two-dimensional electron or hole gas (i.e., the magnetic field is applied along the growth direction), the spin and spatial coordinates do not couple and the respective Hamiltonians can be separated. The eigenvalues of the spin part are given by $\pm \frac{1}{2} g \mu_B B$, while the spatial part will form a harmonic oscillator-like energy distribution $(n + 1/2) \hbar \omega_c$, in addition to the quantum confinement E_L, where g is the g-factor and ω_c is the cyclotron resonance $(\omega_c = \frac{eB}{m^*})$, and L corresponds to the L-th confined level. Consequently, the total energy of each carrier is given by

$$\frac{\gamma_B}{\hbar \omega_c} (2R_0) = (n + \frac{1}{2}) \hbar \omega_c \pm \frac{1}{2} g \mu_B B , \qquad (7.17)$$

where the $\pm$ sign refers to the spin (up or down). For the recombination between Landau levels formed by two-dimensional free electrons and free holes, the mass, m^*, should be replaced by the reduced mass, μ. By fitting the experimental results of the electron-hole transition energies with (7.17), the reduced mass, μ, can be evaluated.

In photoluminescence experiments, measured in the presence of a strong magnetic field applied perpendicular to the interfaces, an oscillatory behavior of the luminescence energy has been demonstrated [224–227]. The magnetic oscillation in the Shubinikov–de Haas measurements of the conductivity occurs due to the spin splitting as a result of the effective g-factor [228]. The narrow linewidth of the cyclotron resonance has been also observed [229] due to the screening of the 2D electron system in a strong magnetic field. There are two factors of importance for the electron energy: The screened exchange interaction and the correlation effects, i.e., the creation of the Coulomb hole or the insufficient electron density surrounding each electron via the virtual density excitations. These two factors are complementary in their role to reduce the self-electron energy. Each of these self-energies exhibits obvious oscillations as a function of the applied magnetic field, but in opposite phase. The dynamical dielectric function, $\epsilon(q, \omega)$, of the 2D electron gas describes both the screening effect on the exchange term and the virtual excitation of the fluctuations in the density. When the Fermi level E_F is at the maximum of the density of states of a Landau level, intra-Landau level excitations will give rise to significant polarization effects. This corresponds to an increase of the real part of the dielectric function, $\mathrm{Re}(\epsilon(q, \omega))$, near $\omega = 0$ as well as the

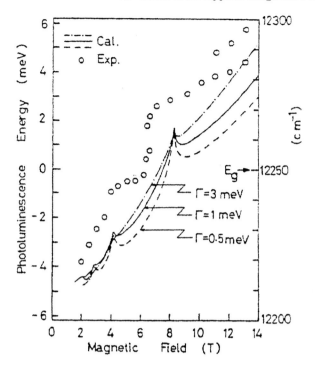

Fig. 7.8. The photoluminescence energy of GaAs/AlGaAs quantum wells with electron density $N_s = 4 \times 10^{11}$ cm^{-2} as a function of the magnetic field. The *dashed, solid,* and *dash-dotted* lines represent the results of calculations for three different Landau-level broadenings, Γ, to be compared with the experimental results (*open circles*) [226]. The lefthand axis is chosen in such a way that zero corresponds to the band gap of bulk GaAs. The right axis shows the absolute value of the photoluminescence energy, expressed in cm^{-1}. (From [227])

imaginary part of Img $(\epsilon (q, \omega))$. These effects will result in an enhancement of the Coulomb-hole term and at the same time a reduction of the exchange term. As the the Fermi level E_F is shifted off the density of states maximum of the Landau level, the intra-Landau level associated polarization decreases and we encounter the opposite situation with a reduction of the Coulomb-hole term and at the same time an increase of the exchange term. Due to the coun-teracting effects of these two terms, there is tendency of cancellation of the oscillation between these contributions and the dependence of the self-energy on the magnetic field decreases. However, such a cancellation effect does not appear for holes, since in this case only the Coulomb hole term is present. This Coulomb term for the holes is similar to the corresponding term for the electrons and exhibits accordingly strong oscillations as the magnetic field is varied. The shift due to the holes is comparable or can even be larger than the corresponding shift for the electrons at higher magnetic fields. Consequently, the net effect on the oscillation of the bandgap renormalization is dominated by the self-energy oscillation of the hole. An example on the oscillation of the photoluminescence energy (*open circles*) with the magnetic field is shown in Fig. 7.8 for GaAs/AlGaAs quantum wells. The authors have performed the calculatations for some different values on the Landau level broadening, Γ. The resulting increase in photoluminescence energy with the increase of Γ is due to the reduced Coulomb-hole term for the hole. As can be seen, a nice agreement between the theoretical predictions and the experimental results are achieved both what concerns the period and the phase of the oscillation.

7.7 Exciton Quenching

The presence of a large number of carriers strongly affects the exciton properties in a heavily doped semiconductor. The principal mechanisms behind the quenching of the excitons are screening, short-range exchange and correlation, and phase-space filling. The quenching process is found to depend on the dimensionality (2D or 3D), doping type (n- or p-type) and the position of the dopant layer (well or barrier doped). In bulk, the screening effect is the major effect and inhibits the exciton formation already at moderate doping levels. For instance, excitons are quenched already at doping levels around $10^{16}\,\mathrm{cm}^{-3}$ in n-type bulk GaAs [230], i.e. well below the metallic limit, while excitons survive all the way up to the degenerate limit ($3 \times 10^{18}\,\mathrm{cm}^{-3}$) in n-type GaAs/AlGaAs quantum well structures [193]. There is a similar trend in p-type structures: Excitons are quenched at $2 \times 10^{17}\,\mathrm{cm}^{-3}$ in p-type bulk GaAs [197], while excitons are found to survive up to hole concentrations as high as $10^{19}\,\mathrm{cm}^{-3}$ (corresponding to a sheet concentration of $1.5 \times 10^{12}\,\mathrm{cm}^{-2}$), i.e. well above the degenerate limit, in quantum wells doped with acceptors in the well [189]. This is illustrated in Fig. 7.9, displaying the photoluminescence spectrum of a quantum well structure with a high doping level ($6 \times 10^{18}\,\mathrm{cm}^{-3}$). Not only the heavy hole state of the free exciton, FE^{hh}, but also the light hole state, FE^{lh}, can be monitored in luminescence. The presence of the FE^{lh} peak with a relatively high luminescence intensity at low temperature (1.6 K) implies that there is a considerable population of light holes at this doping level, i.e., the Fermi-level is close to the light hole subband. Another interesting observation to be made is that the energy separation between the light and heavy hole states of the free exciton, FE^{lh}-FE^{hh}, in luminescence is increasing with increasing doping level, while the opposite behavior is observed in photoluminescence excitation, as illustrated in the insert of Fig. 7.9. In order to verify the interpretation of the FE^{lh} and FE^{hh} peaks, polarisation dependent photoluminescence excitation measurements performed (depicted in the same insert) clearly confirmed the heavy and light hole characteristics, respectively, of the two peaks, as compared with the corresponding unpolarized photoluminescence excitation measurements. This luminescence

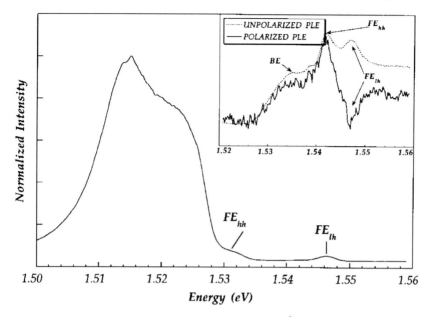

Fig. 7.9. The photoluminescence spectrum of a 150 Å wide GaAs/Al$_{0.3}$Ga$_{0.7}$As quantum well with a doping concentration of 6×10^{18} cm^{-3}, (i.e. above the degenerate limit). A polarized photoluminescence excitation spectrum is depicted in the inset. At such a doping concentration, the light hole free exciton (FE$_{lh}$) is also observed, in addition to the normal heavy hole free exciton (FE$_{hh}$) and the acceptor bound exciton (BE)

excitation spectrum demonstrates clearly that excitons survive also at this doping level, 6×10^{18} cm^{-3}, for the case of center doped quantum wells.

However, the situation differs for the case, when the acceptors are located in the barrier, i.e., p-type modulation doping, instead of in the wells. The acceptor doping level required to quench the excitons is found to be significantly lower than for the case with acceptor doping in the well, but still considerably higher than in the bulk case, due to the inefficiency of screening in a 2D system. The same tendency has earlier been observed in n-type quantum wells: For instance, the excitons were found to be quenched at a sheet concentration of $\approx 4 \times 10^{11}$ cm^{-2} for modulation doped quantum wells [194, 218, 231], while the excitons survived up to a 2D concentration of 1.5×10^{12} cm^{-2} for the center doped wells [193, 196].

The physical background for this significant difference in the quenching behavior is complex, and not fully understood, but there are some important qualitative aspects, which should be taken into consideration: The high doping will give rise to a distortion of the density of states close to the band edges [232]. Due to this fact, the occupancy in k space can be very different for a modulation doped structure in comparison with a center doped one.

Consequently, the prerequisite for formation of excitons will be affected by this distortion. This distortion will also exhibit a different behavior for p-type and n-type quantum wells, due to the difference in binding energy between the acceptors and donors. Another aspect of importance for the quenching of excitons is the screening effect. Reynolds et al. [198] studied the screening effect from a slightly different point of view, but the results are likely of relevance for the topic of interest here. They investigated the dependence of the binding of the donor in the acceptor modulation doped structures as the acceptor location was altered from the center of the well out to the barrier in a co-doped quantum well structure. It was found that the screening effect caused by the free holes from the acceptor is reduced when acceptors are doped inside the well, i.e.due to the presence of the ionized acceptors in the well. This reduced screening effect can be estimated by the increase of the binding energy of the donor, i.e. the donor binding energy is larger in the structure with the acceptors doped in the center region of the well layer than one in the structure with the acceptor doped in the center region of barrier layer. This result implies in a more general sense that the screening effect is influenced by the appearance of ionized acceptors, which in turn would also affect the mechanism of quenching the excitons. Consequently, the filling of the bands will modify the conditions for forming excitons to a more limited extent in the proximity of ionized acceptors in the well in comparison with the case, when the acceptors are located in the barrier due to the screening effect.

7.8 Interacting Impurities

When the impurity concentrations reach such densities that the average spatial separation between impurities become comparable or smaller than the extension of the exciton wave function, there will be an additional contribution to the potential of the impurity binding the exciton. This additive potential contribution is caused by the attractive potential from neighbouring impurities and is naturally dependent on the spatial separation between the interacting impurities. Accordingly, one would expect a distribution of binding energies reflected by the various spatial separations between the interacting impurities, but also possibly dependent on the number of interacting impurities. In general, the larger number of impurities and the smaller interdistance, the larger attractive potential binding the exciton, i.e. the larger binding energy for the exciton.

There are obviously analogues with the so-called undulation spectra, which attracted great interest in the 1970's. The most well known example of the undulation spectra was for the N doped GaP [233, 234], exhibiting a nicely oscillatory structure in the luminescence spectrum. It was claimed that this oscillatory behavior was caused by fluctuations in the number of available impurity pair sites. The undulations could accordingly be seen as the result of averaging the numbers of equivalent sites for each possible pair-

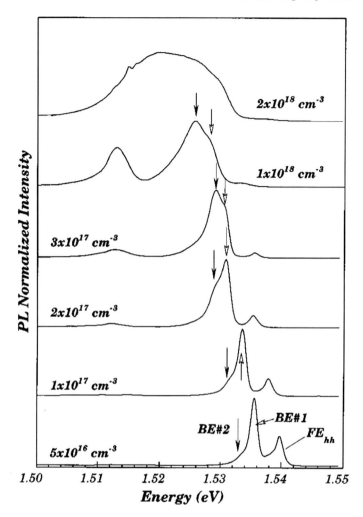

Fig. 7.10. Photoluminescence spectra from acceptor doped GaAs/Al$_{0.3}$Ga$_{0.7}$As quantum wells (similar to the samples shown in Fig. 7.9) at low excitation intensities, with increasing doping concentration. The figure shows the development of BE#2 (*solid arrows*), at the low energy tail of the principal acceptor bound exciton (BE), BE#1 (*open arrows*). The heavy hole state of the free exciton (FE) is also indicated

separation [235, 236]. Later on, Monemar et al. have investigated photoluminescence undulation spectra for acceptors in ZnTe [237]. They concluded that the undulatory structure is not related to the density of states of bound excitons, rather it is a direct spectral manifestation of one-phonon-assisted exciton transfer between neutral acceptors.

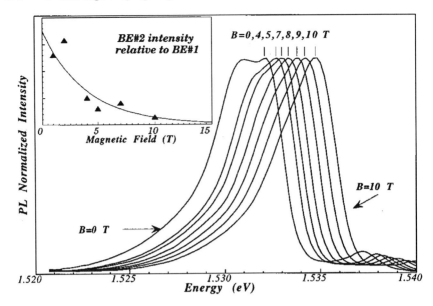

Fig. 7.11. The magnetic field dependence of the photoluminescence spectra of acceptor doped quantum wells with a well width of 150 Å between 0 and 10 T. The inset depicts the relative luminescence intensity of BE#2 relative to BE#1

For the case of quantum wells doped with acceptors in the high doping regime ($> 10^{17}\,\mathrm{cm}^{-3}$), the effect of the interaction between the acceptors can be observed as an additional feature appearing in luminescence spectra on the low energy side of the "conventional" acceptor bound exciton. This is illustrated in Fig. 7.10 by a synopsis of luminescence spectra for a successively increasing acceptor concentration [238]. As seen in this figure, the novel feature appears already at moderate concentrations to gain intensity and become dominant for acceptor concentrations above $3 \times 10^{17}\,\mathrm{cm}^{-3}$. The origin of this feature as being due to the exciton bound at interacting acceptors, is most clearly demonstrated by the magnetic field dependence (Fig. 7.11). As the magnetic field increases, the intensity of the interacting acceptor bound exciton is progressively reduced relatively the normal acceptor bound exciton, due to the shrinkage of the excitonic wave function. Consequently, the probability for the exciton wave function to extend over more than one single acceptor potential decreases.

Upon a closer inspection of the spectra, it can be seen that the binding energy for the interacting acceptor bound exciton, defined as the energy separation relatively the normal acceptor bound exciton, increases from $\approx 1\,\mathrm{meV}$ up to almost $3\,\mathrm{meV}$, as the acceptor concentration is increased by one order of magnitude (to $\approx 10^{18}\,\mathrm{cm}^{-3}$). The intensity ratio between the two acceptor bound excitons has a striking dependence on the excitation intensity. The normal acceptor bound exciton gains intensity versus the interacting acceptor correspondence, as the excitation intensity increases, demonstrating a saturation effect achieved for the interacting acceptor bound excitons. Similarly, the kinetics studies imply that the excitons form a mutually interacting three-level exciton system, resulting in a successively longer decay time, when going from the free exciton to the normal acceptor bound exciton, and further to the interacting acceptor bound exciton.

8 Hydrogen Passivation

The fact that hydrogen interacts efficiently with most defects, shallow dopants as well as deep centers, with the formation of hydrogen-impurity complexes and saturation of dangling bonds as a result, has contributed to a great interest in the hydrogen passivation process. From a technological point of view, the hydrogen passivation plays an important role both at the growth and the processing phases of the semiconductor bulk material. From a fundamental point of view, it is important to reach a thorough understanding of the hydrogen interaction with the impurities. The hydrogen passivation process in quantum wells is significantly less studied than in the case of bulk. One reason for this limited effort is that the driving force for studies of the passivation in bulk, i.e., to improve the radiative recombination efficiency by passivating competing non-radiative processes in as-grown structures constitutes not a similar motivation level in quantum well structures. This fact is due to the considerably higher radiative recombination efficiency in high quality quantum well structures. A comprehensive research has been done on bulk semiconductor materials, particularly on Si. It has been demonstrated that up to 99% of acceptors in Si can be deactivated by atomic hydrogen [239]. In bulk GaAs it is also shown that both donors and acceptors can be passivated by hydrogen. It has been concluded that hydrogen drifts as a positively charged species in p-type GaAs. This means that hydrogen has a donor level in the upper half of the GaAs bandgap. The local vibrational modes of the hydrogen-acceptors, i.e., Be [240, 240–242], Mg [243], Zn [244, 245], C [246], Si_{As} [247], have been observed by far-infrared absorption measurements.

It has been shown that the radiative recombination dominates over the non-radiative processes up to temperatures of approximately 200 K in high quality quantum wells [248]. The nonradiative effects were clearly shown in the temperature dependence of the free exciton decay time. The decay time first increases with increasing temperature up to temperatures of about 200 K, and then decreases with further increasing temperature. The decrease of the life-time was concluded to be due to the increasing importance of nonradiative channels. The presence of the nonradiative recombination at high temperatures was further illustrated by comparing the integrated intensity versus the excitation density at two different temprratures, e.g., at 77 K, where the nonradiative processes are not important and 300 K, where the nonradiative recombination becomes the predominant process. The results clearly show that the integrated luminescence intensity increases faster at 300 K than at 77 K, indicating the presence of a nonradiative channel in addition to the radiative recombination.

There are several techonological applications for hydrogen passivation of shallow and/or deep defects in semiconductors. For instance, the hydrogen passivation is used for improving the carrier mobility and direct control of doping profiles via selective hydrogenation [249, 250]. The role of hydrogen is to link to the unpaired bond of the dopant, sitting either at a bonding position or an anti-bonding site. There are several theoretical approaches on the exact hydrogen-related configuration and the stability of the hydrogen bond [251, 252]. However, while the level of understanding of and knowledge on the hydrogen passivation mechanism is extensive for bulk, the corresponding knowledge level is significantly lower for the case of passivation of defects in quantum wells. Another contributing factor for the limited understanding and also interest is the lower efficiency reported of passivation of impurities in quantum wells [253]. The non-radiative recombination is inefficient in quantum wells at low temperatures already before any passivation [254].

The effect of hydrogen passivation in n-type GaAs/AlGaAs quantum wells was investigated by C.I. Harris et al. [248]. Si shallow donors in GaAs/AlGaAs single quantum well structures were treated by a dc plasma technique. Passivation of the donors was evidenced by the strong changing in the photoluminescence spectra, i.e., disappearing of the Si donor bound excitons line and reappearance of the free excitons component in strongly doped samples. Although dopant impurities are clearly passivated by the hydrogen, there is no evidence to suggest that interface localization is reduced. For high initial doping densities the process of passivation is in fact shown to enhance localization, such that the free exciton transitions are significantly broadened.

Due to limited doped thickness in quantum well structures it is more difficult to perform absorption measurements in such structures. Therefore in quantum well systems, investigations of impurities passivated by atomic hydrogen are very limited. Buyanova et al. [255] reported a photoluminescence study of H-passivated Be-doped AlGaAs/GaAs structures, where they showed that 50% of the Be-acceptors can be deactivated, while the results showed also a strong degradation of the structures after DC hydrogen plasma treatments. The passivation efficiency was found to increase with time of hydrogenation up to a certain level (approx. 1.5 hours) after which a saturation effect is achieved. However, at longer passivation times, also a degradation of the quantum well interfaces was monitored via an increased Stokes shift and broadening of emission spectra. This degradation is explained in terms of an enhanced Ga/Al intermixing.

By comparing the photoluminescence spectra of a series of quantum well structures with various acceptor doping level, I.A. Buyanova et al. [255] showed that there is a strong passivation effect as observed from the spectral blue-shift (about 4 meV) of the exciton spectrum (the free and acceptor bound excitons) in the hydrogen passivated sample in comparison with the

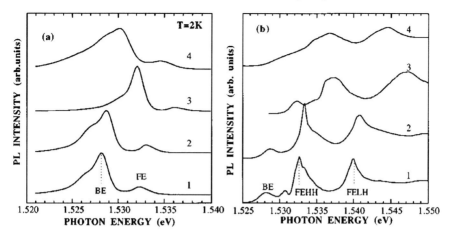

Fig. 8.1. The development of (a) the photoluminescence and (b) photoluminescence excitation spectra with the hydrogen passivation, recorded for GaAs/Al$_{0.3}$Ga$_{0.7}$As quantum wells with acceptor doping level of $2 \times 10^{17}\,\mathrm{cm}^{-3}$. Spectrum 1 corresponds to the as-grown structure, while the spectra 2, 3, and 4 were recorded after the hydrogenation during 40 minutes, 1.5 hours, and 2.5 hours, respectively

corresponding unpassivated structure. The advantage of using the series with various acceptor doping makes it possible to roughly estimate the deactivation level by a simple comparison of the luminescence peak position for the passivated quantum well structure with a similar photoluminescence peak position for another corresponding unpassivated quantum well structure. For the passivated sample with the acceptor concentration $2 \times 10^{17}\,\mathrm{cm}^{-3}$, the deactivation level can accordingly be estimated to be about 50%. For higher doping doping levels, this study of the effect of the hydrogen passivation is less conclusive, since the spectra become too broad.

Another effect of the hydrogen passivation as observed by I.A. Buyanova et al. [255] is on the quantum well interfaces. By investigating the passivation effect on the acceptor doped quantum wells by means of photoluminescence and photoluminescence excitation spectra as a function of the hydrogen passivation time, a blue-shift was observed in the photoluminescence and photoluminescence excitation spectra with increasing passivation time as described above (Fig. 8.1). However, at a closer inspection, one could notice that the blue-shifts observed in photoluminescence and photoluminescence excitation were of different magnitude, with a larger blue-shift observed in photoluminescence excitation, i.e. an increasing Stokes' shift (the energy difference between the heavy hole free exciton positions in photoluminescence and photoluminescence excitation spectra) with increasing passivation time up to about 1 hour. An increasing Stokes' shift is normally an implication of a rougher interface. The interpretation of an increasing degradation of the

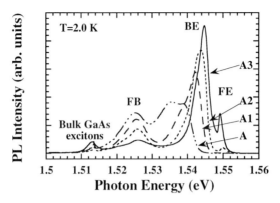

Fig. 8.2. Photoluminescence spectra of an acceptor doped $100\,\text{Å}$ wide $GaAs/Al_{0.3}Ga_{0.7}As$ quantum wells with a doping level of $2 \times 10^{18}\,cm^{-3}$, as-grown structure (A), and different H-treatment (A1-A3), measured at $2.0\,K$ with an excitation wavelength of $5145\,\text{Å}$. The heavy hole free exciton (FE), the acceptor bound exciton (BE) and the transition between free electrons to holes bound at acceptors (FB) are observed

quantum well interface with increasing passivation time is consistent with the observed broadening of the photoluminescence and photoluminescence excitation spectra after additional prolongation (> 1 hour) of the passivation process as demonstrated in Fig. 8.1. The observed degradation is explained in terms of hydrogen-enhanced intermixing at the GaAs/AlGaAs interfaces. It is likely that defects are created due to the presence of hydrogen at the interfaces, which in turn will enhance the Ga diffusion.

The degradation of the quality of quantum well structures has to be avoided if the passivation effects is intended to be used in the device fabrication processed. This has been achieved later on by Q.X. Zhao et al. [256,257]. More systemic studies have been reported for GaAs/AlGaAs [256] and ZnCdTe quantum well structures [258]. They have developed a way to roughly estimate the passivation efficiency by using the photoluminescence technique, and have shown that more than 80% of acceptors in GaAs/AlGaAs quantum well structures can be passivated by hydrogen atoms through hydrogen dc-plasma treatments without causing any degradation of the quantum well structures. The results from a systematic plasma treatment showed that the optimized temperature to achieve high efficiency of hydrogen passivated acceptors in GaAs/AlGaAs quantum well structures by dc-plasma is around $250°C$ [256]. The cooling down conditions are also extremely important to achieve high concentration of H(D)-acceptor complexes, i.e., both the plasma voltage and the bias voltage must be kept on during cooling down process. On the other hand, the duration of plasma treatments is not so critical.

Figure 8.2 illustrates the near bandgap luminescence with different passivation efficiencies [256]. To get some idea about the dependence of the free

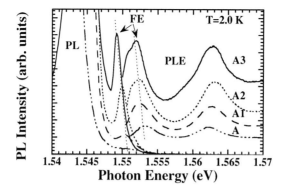

Fig. 8.3. Photoluminescence (PL) and photoluminescence excitation (PLE) spectra of an acceptor doped 100 Å wide GaAs/Al$_{0.3}$Ga$_{0.7}$As quantum well with a doping concentration of 2×10^{18} cm^{-3}, as-grown structure (A), and different H-treatment (A1-A3), measured at 2.0 K. The FE in the figure denotes the heavy hole free exciton. The *solid, dotted, dashed* and *dot-dashed lines* correspond to samples A3 (82% passivation), A2 (75% passivation), A1 (53% passivation) and A (as-grown sample) [257], respectively

exciton localization on the acceptor concentration, the energy positions of the heavy hole free exciton transition measured in the photoluminescence and photoluminescence excitation spectra were compared. The results are summarized in Fig. 8.3 for the 10 nm quantum well structures. The intensity of each spectrum is shifted to make the figure more clear. It can be seen that the shift of the heavy hole free exciton transition in the photoluminescence and photoluminescence excitation spectra is very similar. This indicates that there is no significant change concerning the Stokes shift (energy difference between the heavy hole free exciton positions in photoluminescence and photoluminescence excitation). For the 20 nm quantum well structures, the heavy hole free exciton is hardly observed in the photoluminescence spectrum due to a more efficient capture of carriers by acceptors, in particular for structures containing high acceptor concentration. Nevertheless, the available data also seem to indicate that the heavy hole free exciton Stokes shift is unchanged.

Passivation of acceptor doped multiple quantum wells have also been studied, employing a low energy H implantation with a Kaufmann gun, using sample temperatures of about 300°C [248]. The Be acceptors were partially compensated by this technique, as evidenced by the bound exciton line luminescence intensity compared to the free exciton line. The effect was weaker than for the underlying GaAs buffer layer, and there was evidence for a long term instability in the passivation process: After one year of room temperature storage, the luminescence spectrum from the sample was nearly the same as for the virgin condition before the passivation. Some deep level passivation also occurred, leading to an increase in the measured maximum luminescence lifetime at elevated temperatures.

9 Conclusions

The two facts:

- the presence of a very limited amount of impurities in a semiconductor, will substantially change the conductivity, the optical characteristics as well as other properties and
- the modern and advanced methods for growth of semiconductor quantum structures makes the fabrication of these quantum structures very controlled and accurate

directly imply the importance of the area selected in this book: *Impurities confined in quantum wells*. The effect of shallow impurities on the emission spectra of the quantum structures has been demonstrated several times in this book, but for some applications the effect of the dopant atom itself should be avoided, e.g., due to the increased scattering resulting in a limited charge carrier mobility. In such cases, the dopant layer can be located in the barrier, outside the quantum well, but the desired charge carriers are captured into the well. In any case, the knowledge on the properties of the impurities is very important and constitutes the background for us to review the properties of shallow impurities in quantum well structures in this book. Electronic devices based on these doped quantum wells can be found in manifoldness in our daily life today and the efforts to investigate the impurity properties have been and still are extensive.

In this book, many existing theoretical models and experimental techniques, used to study shallow impurity properties, have been reviewed in detail. One can claim that the understanding of shallow impurity concerning their electronic structures and their optical properties in quantum well structures has today reached a satisfactory level. The electronic structures have been calculated by the effective mass theory, and the results show a good agreement with the experiment findings. In this book, information is provided on shallow donors and acceptors confined a quantum well system, such as their electronic structures, optical properties, dynamics and effective g-factors. Experimental aspects as well as theoretical predictions have been presented. The excess carrier effects due to shallow doping on the optical properties of quantum well structures, impurity interaction and hydrogen passivation have also been briefly discussed. The extensive knowledge of shallow impurity confined in two-dimensional structures and bulk materials constitutes a solid base to further explore impurity properties in systems with a lower dimensionality; quantum wires and dots.

Despite the extensive knowledge on shallow impurities in quantum well systems, there are some essential problems remaining to be solved: For instance, the intra-subband transitions of impurity levels are not well experimentally documented in quantum well structures, due to the difficulty of the far infrared absorption measurements in quantum well structures. Therefore, the theoretical predictions still remain to be verified, particularly in the presence of an external magnetic field or an external pressure. Furthermore, the investigations on the dynamic properties concerning the intra-impurity-level relaxation is very limited, and the decrease of the relaxation time between the acceptor 2p–1s states in quantum well structures in comparison with bulk case is not well understood.

We try to include as much as possible of the literature reports in this field. However, we may still miss some original works. In this case, we apologize to the authors.

References

1. G. Bastard, J.A. Brum, and R. Ferreira, *Electronic States in Semiconductor Heterostructures, Solid State Physics, Advances in Research and Applications,* Vol. 44, (Academic Press, Boston 1991)
2. C. Weisbuch and B. Winter, *Quantum Semiconductor Structures,* (Academic Press, Boston 1991)
3. P.Y. Yu and M. Cardona, *Fundamentals of Semiconductors,* (Springer Verlag, Berlin 1996)
4. C. Weisbuch, Applications of multiquantum wells, selective doping, and superlattices, In: *Semiconductors and Semimetals,* Vol.24 , (Academic Press Inc. Ltd, London, 1987)
5. G.E. Stillman and C.M. Wolfe, In: *Semiconductors and Semimetals,* Vol.12, ed. by R.K. Willardson and A.C. Beer, (Academic, New York, 1977) pp.169
6. T.D. Harris, M.S. Skolnick, J.M. Parsey Jr, and R. Bhat, Appl. Phys. Lett. **52**, 389 (1988)
7. V.A. Karasyuk, D.G.S. Beckett, M.K. Nissen , A. Villemaire, T.W. Steiner, and M.L.W. Thewalt, Phys. Rev. **B49**, 16381 (1994)
8. C. Weibuch and C. Hermann, Phys. Rev. **B15**, 816 (1976)
9. E.O. Kane, Phys. Rev. Lett. **12**, 97-98 (1964)
10. E. Cohen, M.D. Sturge, W.O. Lipari, M. Altarelli, and A. Baldereshi, Phys. Rev. Lett. **35**, 1591 (1975)
11. Q.X. Zhao and B. Monemar, Phys. Rev. **B3**, 1397 (1988)
12. Q.X. Zhao, P. Bergman, and B. Monemar, Phys. Rev. **B38**, 8383 (1988)
13. B. Monemar, P.O. Holtz, W.M. Chen, H.P. Gislason, U. Lindefelt, and M.E. Pistol, Phys. Rev. **B34**, 8656 (1986)
14. W.M. Chen, Q.X. Zhao, B. Monemar, H.P. Gislason, and P.O. Holtz, Phys. Rev. **B33**, 3722 (1987)
15. N. Killoran, D.J. Dunstan, M.O. Henry, E.C. Lightowlers, and B.C. Cavenett, J. Phys. **C15**, 6067 (1982)
16. P.O. Holtz, Q.X. Zhao, and B. Monemar, Phys. Rev. **B36**, 5051 (1987)
17. J.P. Hopfield, R.G. Thomas, and R.T. Lynch, Phys. Rev. Lett. **17**, 312 (1966)
18. B. Monemar, U. Lindefelt, and W.M. Chen, Physica B + C **146B**, 256 (1987)
19. Q.X. Zhao and T. Westgaard, Phys. Rev. **B44**, 3726 (1991)
20. Q.X. Zhao and B. Monemar, Phys. Rev. **B38**, 1397 (1988)
21. G. Bastard, Phys. Rev. **B24**, 4714 (1981)
22. C. Mailhiot, Y.C. Chang, and T.C. McGill, Journal Vac. Sci. Technology **21**, 519 (1982)
23. C. Mailhiot, Y.C. Chang, and T.C. McGill, Phys. Rev. **B26**, 4449 (1982)
24. R.L. Greene and K.K. Bajaj, Solid State Commun. **45**, 825 (1983)

25. S. Fraizzoli and A. Pasquarello, Phys. Rev. **B44**, 1118 (1991)
26. S. Fraizzoli, F. Bassani, and R. Buczko, Phys. Rev. **B41**, 5096 (1990)
27. C. Mailhiot, Y.C. Chang, and T.C. McGill, Journal Vac. Sci. Technology **21**, 519 (1982)
28. C. Mailhiot, Y.C. Chang, and T.C. McGill, Phys. Rev. **B26**, 4449 (1982)
29. M. Stopa and S. DasSarma, Phys. Rev. **B40**, 8466 (1989)
30. R.L. Greene and K.K. Bajaj, Phys. Rev. **B31**, 913 (1985)
31. R.L. Greene and K.K. Bajaj, Phys. Rev. **B31**, 4006 (1985)
32. G. Bastard, E.E. Mendez, L.L. Chang, and L. Esaki, Phys. Rev. **B28**, 3241 (1983)
33. D. Ahn and Chuang, Phys. Rev. B34, 9034 (1986)
34. B. Sunder, Phys. Rev. B45, 8562 (1992)
35. C. Alibert, S. Gaillard, J.A. Brum, G. Bastard, P. Frijlink, and M. Erman, Solid State Commun. **53**, 457 (1985)
36. P.C. Klipstein, P.R. Tapster, N. Aspley, D.A. Anderson, M.S. Skolnick, T.M. Kerr, and K. Woodbrigde, J. Phys. C**19**, 857 (1985)
37. M. Matsuura and T. Kamizato, Phys. Rev. **B33**, 8385 (1986)
38. A. Petrou, J. Warnock, J. Ralston, and G. Wicks, Solid State Commun. **58**, 581 (1986)
39. S. Yoo, L. He, B.D. McCombe, and W.D. Schaff, Superlattices and Microstructures Vol. **8**, No 3, 297 (1990)
40. S. Huant, W. Knap, G. Martinez, and B. Etienne, Europhysics Letters **7**, 159 (1988)
41. J.M. Mercy, N.C. Jarosik, B.D. McCombe, J. Ralston, and G. Wicks, Journal of Vacuum and Science Technology, B **4**, 1011 (1986)
42. S. Chaudhuri and K.K. Bajaj, Solid State Commun. **52**, 967 (1984)
43. D.M. Larsen, Phys. Rev. **B 44**, 5629 (1991)
44. L.H.M. Barbosa, A. Latge, M. de Dios Leyva, and L.E. Oliveira, Solid State Commun. **98**, 215 (1996)
45. Y.T. Yip and W.C. Kok, Phys. Rev. **B 59**, 15825 (1999)
46. J. Cen, S.M. Lee, K.K. Bajaj, J. Appl. Phys. **73**, 2848 81993)
47. P.W. Barmby, J.L. Dunn, and C.A. Bates, Journal of Phys. C **7**, 2473 (1995)
48. N.C. Jarosik, B.D. McCombe, B.V. Shanabrook, J. Comas, J. Ralston, and G. Wicks, Phys. Rev. Lett. **54**, 1283 (1985)
49. A.A. Reeder, J.M. Mercy, and B.D. McCombe, IEEE Journal of Quantum Electronics, Vol. **24**, 1690 (1988)
50. J.P. Cheng and B.D. McCombe, Phys. Rev. **B42**, 7626 (1990)
51. B.V. Shanabrook, J. Comas, T.A. Perry, R. Merlin, Phys. Rev. B29, 7096 (1984)
52. D. Wolverson, S.V. Railson, M.P. Halsall, J.J. Davies, D.E. Ashenford, B. Lunn, Semiconductor Science and Technology **10**, 1475 (1995)
53. M.P. Halsall, S.V. Railson, D. Wolverson, J.J. Davies, B. Lunn, and D.E. Ashenford, Phys. Rev. **B 50**, 11755 (1994)
54. M.P. Halsall, D. Wolverson, S J.J. Davies, D.E. Ashenford, and B. Lunn, Solid State Communications **86**, 15 (1993)
55. R. Mayer, M. Dahl, G. Schaack, and A. Waag, Phys. Rev. **B 55**, 16376 (1997)
56. I.I. Reshina, S.V. Ivanov, D.N. Mirlin, A.A. Toropov, A. Waag, and G. Landwehr, Phys. Stat. Sol. (b) **229**, 685 (2002)
57. W.E. Hagston, P. Harrison, and T. Stirner, Phys. Rev. **B 49**, 8242 (1994)

58. J.P. Bergman, P.O. Holtz, B. Monemar, M. Sundaram, J.L. Merz, and A.C. Gossard, Phys. Rev. **B 43**, 4765 (1991)

59. J. Feldmann, M. Preis, E.O. Gobel, P. Dawson, C.T. Foxon, and I. Galbraith, Solid State Commun. **83**, 245 (1992)

60. B.V. Shanabrook and J. Comas, Surf. Sci. **142**, 504 (1984)

61. B.V. Shanabrook, Surf. Sci. **170**, 449 (1986)

62. M.A. Lampert, Phys. Rev. Lett. **1**, 450 (1958)

63. J.R. Haynes, Phys. Rev. Lett. **4**, 361 (1960)

64. J.J. Hopfield, Proc. of the International Conference of the Physics of Semiconductors, Paris, 1964, ed. by Dunod, p. 725

65. T. Skettrup, M. Suffczynski, and W. Gorzkowski, Phys. Rev. **B4**, 512 (1971)

66. D.G. Thomas and J.J. Hopfield, Phys. Rev. **128**, 2135 (1962)

67. T.D. Harris, M.S. Skolnick, J.M. Parsey, and R. Bhat, Appl. Phys. Lett. **52**, 389 (1988)

68. J.M. Luttinger, Phys. Rev. **102**, 1030 (1955)

69. D. Bimberg and P.J. Dean, Phys. Rev. **15**, 3917 (1976)

70. B.V. Shanabrook and J. Comas, Surf. Sci. **142**, 504 (1984)

71. B.V. Shanabrook, Surf. Sci. **170**, 449 (1986)

72. Y. Nomura, K. Shinozaki, and M. Ishii, Journal of Appl. Phys. **58**, 1864 (1985)

73. S. Charbonneu, T. Steiner, M.L.W. Thewalt, E.S. Koteles, J.Y. Chi, and B. Elman, Phys. Rev. **B38**, 3583(1988)

74. X. Liu, A. Petrou, B.D. McCombe, J. Ralston, and G. Wicks, Phys. Rev. **B38**, 8522 (1988)

75. D.C. Reynolds, K.G. Merkel, C.E Schutz, K.R Eaves, and P.W. Yu, Phys. Rev. **B43**, 1604 (1991)

76. D.C. Reynolds, K.R Eaves, K.G. Merkel, C.E Schutz, and P.W. Yu, Phys. Rev. **B43**, 9087 (1991)

77. D.C. Reynolds, C.E. Leak, K.K. Bajaj, C.E Schutz, R.L. Jones, K.R Eaves, P.W. Yu, and W.M. Theis, Phys. Rev. **B40**, 6210 (1989)

78. B.V. Shanabrook, J. Comas, T.A. Perry, and R. Merlin, Phys. Rev. **B29**, 7096 (1984)

79. B.V. Shanabrook and J. Comas, Surf. Sci. **142**, 504 (1984)

80. Y. Nomura, K. Shinozaki, and M. Ishii, J. Appl. Phys. **58**, 1864 (1985)

81. T.D. Harris, M.S. Skolnick, J.M. Parsey Jr, and R. Bhat, Appl. Phys. Lett. **52**, 389, (1988)

82. P.O. Holtz, B. Monemar, M. Sundaram, J.L. Merz, and A.C. Gossard, Phys. Rev. B 47, 10596 (1993)

83. A. MacDonald and D. Ritchie, Phys. Rev. B33, 8336 (1986)

84. J.-P. Cheng and B.D. McCombe, Phys. Rev. **B42**, 7626 (1990)

85. A. MacDonald and D. Ritchie, Phys. Rev. **B33**, 8336 (1986)

86. R.C. Miller, A.C. Gossard, and D.A. Kleinman, Phys. Rev. **B 32**, 5443-5446 (1985)

87. J.L. Dunn, E.Pearl, R.T. Grimes, M.B. Stanaway, and J.M. Chamberlain, Materials Science Forum Vols. **65 - 66**, 117 (1990)

88. R.T. Grimes, M.B. Stanaway, J.M. Chamberlain, J.L. Dunn, M. Henini, O.H. Hughes, and G. Hill, Semiconductor Science Technology, Vol. **5**, 305 (1990)

89. C.J. Armistead, R.A. Stradling, and Z. Wasilewski, Semiconductor Science Technology, Vol. **4**, 557 (1989)

90. D.C. Reynolds, K.R. Evans, C.E. Stutz, and P.W. Yu, Phys. Rev. **B 44**, 1839 (1991)

91. G.A. Balchin, L.M. Smith, A. Petrou, and B.D. McCombe, Superlattices and Microstructures, **18**, 291 (1995)

92. B. Monemar, H. Kalt, C. Harris, J.P. Bergman, P.O. Holtz, M. Sundaram, J.L. Merz, A.C Gossard, K. Köhler, T. Schweizer, Superlattices and Microstructures, **9**, 281 (1991)

93. M. O'Neill, P. Harrison, M. Oestreich, D.E. Ashenford, J. Appl. Phys. **78**, 451 (1995)

94. T. Pang and S.G. Louie, Phys. Rev. Lett. **65**, 1635 (1990)

95. E.R. Mueller, D.M. Larsen, J. Waldman, and W.D. Goodhue, Phys. Rev. Lett. **68**, 2204 (1992)

96. D.M. Larsen and S.Y. McCann, Phys. Rev. **B45**, 3485 (1992)

97. S. Huant, S.P. Najda, and B. Etienne, Phys. Rev. Lett. **65**, 1486 (1990)

98. S. Huant, A. Mandray, G. Martinez, M. Grynberg, and B. Etienne, Surf. Sci. **263**, 565 (1992)

99. C.J. Armistead, S.P. Najda, R.A. Stradling, and J.C. Maan, Solid State Commun. **53**, 1109 (1985)

100. S. Huant, S.P. Najda, and B. Etienne, Phys. Rev. Lett. **65**, 1486 (1990)

101. J.P. Cheng, Y.J. Wang, B.D. McCombe, and W. Schaff, Phys. Rev. Lett. **70**, 489 (1993)

102. P. Hawrylak, Phys. Rev. Lett. **72**, 2943 (1994)

103. J. Blinowski and T. Szwacka, Phys. Rev. **B49**, 10231 (1994)

104. W.J. Li, J.L. Wang, B.D. McCombe, J.P. Cheng, and W. Schaff, Surface Science, Vol. **305**, 215 (1994)

105. A.B. Dzyubenko, A. Mandray, S. Huant, A.Y. Sivachenko, and B. Etienne, Phys. Rev. **B50**, 4687 (1994)

106. J. Kono, S.T. Lee, M.S. Salib, G.S. Herold, A. Petrou, and B.D. McCombe, Phys. Rev. **B52**, R8654 (1995)

107. J.M. Luttinger and W. Kohn, Phys. Rev. **97**, 869 (1955)

108. W.T. Masselink, Y.-C. Change, and H. Morkoç, Phys. Rev. **B28**, 7373 (1983); B32, 5190 (1985)

109. J.P. Loehr, Y.C. Chen, D. Biswas, P.K. Bhattacharya, and J. Singh, Proceeding of the 20th international conference on the physics of semiconductors (World Scientific, Singapore, 1990), Vol. **2**, p. 1404

110. S. Fraizzoli, and A. Pasquarello, Phys. Rev. **B42**, 5349 (1990); **B44**, 1118 (1991)

111. A. Pasquarello, L.C. Andreani, and R. Buczko, Phys. Rev. **B40**, 5602 (1989)

112. G. Einevoll and Y.C. Chang, Phys. Rev. **B41**, 1447 (1990)

113. Q.X. Zhao, A. Pasquarello, P.O. Holtz, B. Monemar, and M. Willander, Phys. Rev. **B50**, 10953 (1994)

114. M. Cai, W. Liu, and Y. Liu, Phys. Rev. **B 46**, 4281 (1992)

115. Q.X. Zhao and M. Willander, Phys. Rev. **B57**, 13033 (1998); J.Appl. Phys. **86**, 5624 (1999)

116. Q.X. Zhao, P.O. Holtz, A. Pasquarello, B. Monemar, and M. Willander, Phys. Rev. **B50**, 2393 (1994)

117. Q.X. Zhao, P.O. Holtz, A. Pasquarello, B. Monemar, A.C. Ferreira, M. Sundaram, J.L. Merz, and A.C. Gossard, Phys. Rev. **B 49**, 10794 (1994)

118. R.F. Kirkman, R.A. Stradling, and P.J. Lin-Chung, J. Phys. C. **11**, 419 (1987)

119. R.C. Miller, A.C. Gossard, W.T. Tsang, and O. Munteanu, Phys. Rev. **B25**, 3871 (1982)

120. A.A. Reeder, B.D. McCombe, F.A. Chambers, and G.P. Devane, Phys. Rev. **B38**, 4318 (1988)

121. B.V. Shanabrook, J. Comas, T.A. Perry, and R. Merlin, Phys. Rev. **B 29**, 7096 (1984)

122. T.A. Perry, R. Merlin, B.V. Shanabrook, and J. Comas, Phys. Rev. Lett. **54**, 2623 (1985)

123. G. Abstreiter, M. Cardona, and A. Pinczuk, in: *Light Scattering in Solids IV*, ed. by M. Cardona and G. Guntherodt, Topics in Applied Physics, Vol. **54** (Springer, Berlin 1984), p. 81

124. V.F. Sapega, M. Cardona, K. Ploog, E.L. Ivchenko, and D.N. Mirlin, Phys. Rev. B **45**, 4320 (1992)

125. H.W. van Kesteren, E.C. Cosman, W.A.J.A. van der Pole, and C.T. Foxon, Phys. Rev. **B 41**, 5283 (1990)

126. D.N. Mirlin and A.A. Sirenko, Sov. Phys. Solid State **34**, 108 (1992)

127. R. Ebert, H.Pascher, G. Appold, and H.G. Hafele, Appl. Phys. **14**, 155 (1977)

128. J.F. Scott, Phys. Rep. 194, 379 (1990); Phys. Rev. Lett. **44**, 1358 (1980)

129. R. Bauer, D. Bimberg, J. Christen, D. Oertel, D. Mars, J.N. Miller, T. Fukunaga, and H. Nakashima, in: *Proceedings of the 18th International Conference on the Physics of Semiconductors*, Stockholm, 1986, ed. by O. Engström (World Scietific, Singapore, 1987), p. 525

130. Y. Chen, B. Gil, P. Lefebre, and H. Mathieu, Phys. Rev. **B 37**, 6429 (1988)

131. W.T. Masselink, P.J. Pearah, J. Klem, C.K. Peng, H. Morkoç, G.D. Sanders, and Y.C. Chang, Phys. Rev. **B32**, 8027 (1985)

132. S.Charbonneau, T. Steiner, M.L.W. Thewalt, E.S. Koteles, J.Y. Chi, and B. Elman, Phys. Rev. **B38**, 3583 (1988)

133. R.T. Phillips, D.J. Lovering, G.J. Denton, and G.W. Smith, Phys. Rev. **B45**, 4308 (1992)

134. J.P. Bergman, P.O. Holtz, B. Monemar, M. Sundaram, J.L. Merz, and A.C. Gossard, in: *Proc. of the International Meeting on Optics of Excitons in Confined Systems*, Inst. Phys. Conf. Ser. No 123, p. 73, 1991

135. R.C. Miller, D.A. Kleinman, W.A. Nordland Jr, and A.C. Gossard, Phys. Rev. **B22**, 863 (1980)

136. C. Weisbuch, R.C. Miller, R. Dingle, A.C. Gossard, and W.Wiegmann, Solid State Commun. **37**, 219 (1981)

137. P.M. Petroff, C. Weisbuch, R. Dingle, A.C. Gossard, and W. Wiegmann, Appl. Phys. Lett. **40**, 507 (1982)

138. P.O. Holtz, Q.X. Zhao, A.C. Ferreira, B. Monemar, M. Sundaram, J.L. Merz, and A.C. Gossard, Phys. Rev. **B48**, 8872 (1993)

139. G.C. Rune, P.O. Holtz, B. Monemar, M. Sundaram, J.L. Merz, and A.C. Gossard, Phys. Rev. **B44**, 4010 (1991) and references therein

140. R.C. Miller, A.C. Gossard, W.T. Tsang, and O. Munteanu, Phys. Rev. **B25**, 3871 (1982)

141. K. Muraki, Y. Takahashi, A. Fujiwara, S. Fukatsu, and Y. Shiraki, Solid State Electronics, Vol. **37**, 1247 (1994)

142. J.A. Brum and G. Bastard, Phys. Rev. **B33**, 1420 (1986)

143. A. Fujiwara, S. Fukatsu, Y. Shiraki, and R. Ito, Surface Science **263**, 642 (1992)

144. P.W.M. Blom, C. Smit, J.E.M. Haverkort, and J.H. Wolter, Phys. Rev. **B47**, 2072 (1993)

145. D. Morris, B. Deveaud, A. Regreny, and P. Auvray, Phys. Rev. **B47**, 6819 (1993)

146. P.O. Holtz, M. Sundaram, J.L. Merz, and A.C. Gossard, Phys. Rev. **B41**, 1489 (1991)

147. A.C. Ferreira, P.O. Holtz, B. Monemar, M. Sundaram, J.L. Merz, and A.C. Gossard, Appl. Phys. Lett. **65**, 720 (1994)

148. M. Zachau, J.A. Kash, and W.T. Masselink, Phys. Rev. **B44**, 4048 (1991)

149. J.A. Kash, M. Zachau, M.A. Tischler, and U. Ekenberg, Phys. Rev. Lett. **69**, 2260 (1992)

150. J.A. Kash, M. Zachau, M.A. Tischler, and U. Ekenberg, Surface Science **305**, 251 (1994)

151. A.M. White, P.J. Dean, and B. Day, Journal of Phys. C **7**, 1400 (1974)

152. M. Lampert, Phys. Rev. Lett. **1**, 450 (1958)

153. P.J. Dean, J.D. Cuthbert, D.G. Thomas, and R.T. Lynch, Phys. Rev. Lett. **18**, 122 (1967)

154. P.J. Dean and D.C. Herbert, in: *Topics in Current Physics*, Vol. 14, Excitons, ed. by K. Cho (Springer Verlag, 1979) pp. 55 and references therein

155. P.J. Dean, H. Venghaus, J.C. Pfister, B. Schaub, and J. Marine, Journal of Lumin. **16**, 363 (1978)

156. P.J. Dean, J.R. Haynes, and W.F. Flood, Phys. Rev. **161**, 711 (1967)

157. P.O. Holtz, Q.X. Zhao, B. Monemar, M. Sundaram, J.L. Merz, and A.C. Gossard, Phys. Rev. **B47**, 15675 (1993)

158. W.T. Masselink, Y.-C. Chang, H. Morkoç, D.C. Reynolds, C W. Litton, K.K. Bajaj, and P.W. Yu, Solid State Electr. **29**, 205 (1986)

159. J.C. Garcia, A.C. Beye, J.P. Contour, G. Neu, J. Massies and A. Barski, Appl. Phys. Lett. **52**, 1596 (1988)

160. R.C. Miller, J. Appl. Phys. **56**, 1136 (1984)

161. P.Y. Yu, Phys. Rev. **B 20**, 5286 (1979)

162. P.O. Holtz, Q.X. Zhao, A.C. Ferreira, B. Monemar, A. Pasquarello, M. Sundaram, J.L. Merz, and A.C. Gossard, Phys. Rev. **B50**, 4901 (1994)

163. M. Altarelli and N.O. Lipari, Phys. Rev. **B7**, 3798 (1973)

164. Q.X. Zhao, M. Karlsteen, M. Willander, S.M. Wang, and M. Sadeghi, Phys. Rev. **B62**, 5055 (2000)

165. J. Feldman, G. Peter, E.O. Göbel, P. Dawson, K. Moore, C. Foxon, and R.J. Elliott, Phys. Rev. Lett. **59**, 2337 (1987)

166. E.O. Göbel, H. Jung, J. Kuhl, and K. Ploog, Phys. Rev. Lett. **51**, 1588 (1983)

167. J. Christen, D. Bimberg, A. Steckenborn, G. Weimann, and W. Sclapp, Superlattices and Microstructures **2**, 251 (1986)

168. H. Stolz, D. Schwarze, W. von der Osten, and G. Weiman, Superlattices and Microstructures **6**, 271 (1989)

169. G. Bastard, C. Delalande, M.H. Meynadier, P.M. Frijlink, and M. Voos, Phys. Rev. **B 29**, 7042 (1984)

170. T. Amand, F. Lephay, S. Valloggia, F. Voillot, M. Brousseau, and A. Regreny, Superlattices and Microstructures **6**, 323 (1989)

171. B. Deveaud, T.C. Damen, J. Shah, and C.W. Tu, Appl. Phys. Lett. **51**, 828 (1987)

172. G.W. Hooft, W.A.J.A. van der Poel, L.W. Molenkamp, and C.T. Foxon, Phys. Rev. **B 35**, 8281 (1987)

173. J.M. Rorison and D.C. Herbert, Superlattices and Microstructures **1**, 423 (1985)
174. E. Finkman, M.D. Sturge, and R. Bhut, J. Luminescence **35**, 235 (1986)
175. Q.X. Zhao, P.O. Holtz, C.I. Harris, B. Monemar, and E. Veje, Appl. Phys. Lett. **64**, 2721, (1994)
176. Q.X. Zhao, P.O. Holtz, C.I. Harris, B. Monemar, and E. Veje, Phys. Rev. **B50**, 2023 (1994)
177. M.P. Halsall, P. Harrison, J.P.R. Wells, I.V. Bradley, and H. Pelleman, Phys. Rev. **B 63**, 155314 (2001)
178. J. Serre, A. Ghazali, and A. Gold, Phys. Rev. **B 39**, 8499 (1989)
179. N.F. Mott, in: *Metal-Insulator Transitions* (Taylor and Francis, London, 1974)
180. G.D. Mahan, Phys. Rev. **153**, 882 (1967)
181. S. Schmitt-Rink, C. Ell, and H. Haug, Phys. Rev. **B33**, 1183 (1986)
182. M.S. Skolnick, J.M. Rorison, K.J. Nash, D.J. Mowbray, P.R. Tapster, S.J. Bass, and A.D. Pitt, Phys. Rev. Lett. **58**, 2130 (1987)
183. W. Chen, M. Fritze, A.V. Numikko, M. Hong, and L.L. Chang, Phys. Rev. **B 43**, 14738 (1991)
184. J.F. Mueller, Phys. Rev. **B 42**, 11189 (1990)
185. B.E. Sernelius, Phys. Rev. **B33**, 8582 (1986)
186. B.E. Sernelius, Phys. Rev. **B34**, 5610 (1986)
187. J.M. Luttinger, Phys. Rev. **102**, 1030 (1956)
188. U. Ekenberg and M. Altarelli, Phys. Rev. **B 32**, 3712 (1985)
189. A.C. Ferreira, P.O. Holtz, B.E. Sernelius, I. Buyanova, B. Monemar, O. Mauritz, U. Ekenberg, M. Sundaram, K. Campman, J.L. Merz, and A.C. Gossard, Phys. Rev. **B 54**, 16989 (1996)
190. P.O. Holtz, A.C. Ferreira, B.E. Sernelius, A. Buyanov, B. Monemar, O. Mauritz, U. Ekenberg, M. Sundaram, K. Campman, J.L. Merz, and A.C. Gossard, Phys. Rev. **B 58**, 4624 (1998)
191. S. Schmitt-Rink, D.S. Chemla, and D.A.B. Miller, Adv. Phys. **38**, 89 (1989)
192. G. Tränkle, H. Leider, A. Forchel, H. Huang, C. Ell, and G. Weimann, Phys. Rev. Lett. **58**, 419 (1987)
193. C.I. Harris, B. Monemar, H. Kalt, K. Köhler, Phys. Rev. **B 48**, 4687 (1993)
194. D. Huang, H.Y. Chu, Y.C. Chang, R. Houdre, and H. Morkoç, Phys. Rev. **B 38**, 1246 (1988)
195. S. Schmitt-Raink, D.S. Chemla, and D.A.B. Miller, Phys. Rev. **B 32**, 6601 (1985)
196. C.I. Harris, H. Kalt, B. Monemar, and K. Köhler, Surf. Sci. **263**, 462 (1991)
197. D. Olego, M. Cardona, Phys. Rev. **B 22**, 886 (1980)
198. D.C. Reynolds, B. Jogai, P.W. Yu, K.R. Evans, and C.E. Stutz, Phys. Rev. **B 46**, 15274 (1992)
199. D.A. Cusano, Solid State Commun. **2**, 353 (1964)
200. D.E. Hill, Phys. Rev. 133, 866 (1963)
201. G. Borghs, K. Bhattacharyya, K. Deneffe, P. Van Mieghem, and R. Mertens, J. Appl. Phys. **66**, 4381 (1989)
202. I. Pankove, J. Appl. Phys. **39**, 5368 (1968)
203. D.A. Broido and L.J. Sham, Phys. Rev. **B31**, 888 (1985)
204. M. Altarelli, Phys. Rev. **B28**, 842 (1983)
205. O. Betbeder-Matibet, M. Combescot, and C. Tanguy, Phys. Rev. Lett. **72**, 4125 (1994)

206. T. Kawabata, K. Muro, and S. Narita, Solid State Commun. **23**, 267 (1977)
207. B. Stébé et al., Solid State Commun. **26**, 637 (1978)
208. G. Munschy and B. Stébé, Phys. Status Solidi (b) **64**, 213 (1974)
209. B. Stébé and A. Ainane, Superlattices and Microstructures **5**, 545 (1989)
210. K. Kheng, R.T. Cox, Y. Merle d'Aubigine, F. Bassani, K. Saminadayar, and S. Tatarenko, Phys. Rev. Lett. **71**, 1752 (1993)
211. G. Finkelstein et al., Phys. Rev. Lett. **74**, 976 (1995)
212. A.J. Shields et al., Phys. Rev. **B51**, 17128 (1995)
213. H.W. Yoon, D.R. Wake, and J.P. Wolfe, Phys. Rev. **B51**, 2763 (1996)
214. A.J. Shields, M. Pepper, D.A. Ritchie, M.Y. Simmons, and G.A.C. Jones, Phys. Rev. **B51**, 18049 (1995)
215. A.J. Shields, M. Pepper, D.A. Ritchie, M.Y. Simmons, and G.A.C. Jones, Phys. Rev. **B52**, 7841 (1995)
216. A.J. Shields, F.M. Bolton, M.Y. Simmons, M. Pepper, D.A. Ritchie, Phys. Rev. **B55**, R1970 (1997)
217. G. Finkelstein, H. Shtrikman, and I. Bar-Joseph, Phys. Rev. **B53**, 12593 (1996)
218. M. Potemski, R. Stepniewski, J.C. Maan, G. Martinez, P. Wyder, and B. Etienne, Phys. Rev. Lett. **66**, 2239 (1991)
219. W. Ossau et al., In: *Properties of impurity States in Superlattice Semiconductors*, Vol. 183 of NATO Advanced Study Institute, Series B: Physics, ed. by E.Y. Fong et al. (Plenum, New York, 1988), p. 285
220. M.J. Snelling et al., Phys. Rev. **B45**, 3922 (1992)
221. A.J. Shields, J.L. Osborne, M.Y. Simmons, M.Pepper, and D.A. Ritchie, Phys. Rev. **B52**, R5523 (1995)
222. J.L. Osborne, A.J. Shields, M. Pepper, F.M. Bolton, and D.A. Ritchie, Phys. Rev. **B53**, 13002 (1996)
223. J.C. Maan, In: *Physics and Applications of Quantum Wells and Superlattices*, ed. by E.E. Mendez and K. von Klitzing, ASI series B: Physics (Plenum, New York, 1987), Vol. 170, p. 347
224. J.M. Worlock, A.C. Maciel, A. Petrou, C.H. Perry, R.L. Aggrawal, M.C. Smith, A.C. Gossard, and W. Wiegmann, Surf. Sci. **142**, 486 (1984)
225. M.C. Smith, A. Petrou, C.H. Perry, J.M. Worlock, and R.L. Aggrawal, In: *Proc. 17th Int'l Conf. Phys. Semicond.*, p. 547 San Francisco (1984)
226. C.H. Perry, J.M. Worlock, M.C. Smith, and A. Petrou, In: *Proc. Int'l Conf. High Magn. in Semicond.Phys.*, p. 202 Springer (1986)
227. S. Katamaya and T. Ando, Solid State Commun. Vol. **70**, 97 (1989)
228. R.J. Nicholas, R.J. Haug, K. von Klitzing, and G. Weimann, Phys. Rev. **B37**, 1294 (1988)
229. Z. Schlesinger, W.I. Wang, and A.H. MacDonald, Phys. Rev. Lett. **58**, 73 (1987)
230. J. Shah, R.F. Leheny, and W. Wiegmann, Phys. Rev. **B16**, 1577 (1977)
231. R. Stepniewski, W. Knap, A. Raymond, G. Martinez, J.C. Maan, and B. Etienne, Surf. Science, **229**, 519 (1990)
232. A. Gold, A. Ghazali, and J. Serre, Phys. Rev. **B40**, 5806 (1989)
233. J.J. Hopfield, H. Kukumoto, and P.J. Dean, Phys. Rev. Lett. **27**, 139 (1971)
234. T.N. Morgan, M.R. Lorenz, and A. Onton, Phys. Rev. Lett. **28**, 906 (1972)
235. R.A Street and P.J. Wiesner, Phys. Rev. Lett. **34**, 1569 (1975) and Phys. Rev. **B14**, 632 (1976)

236. P.J. Wiesner, R.A Street, and H.D. Wolf, Phys. Rev. Lett. **35**, 1366 (1975)
237. B. Monemar, N. Magnea, and P.O. Holtz, Phys. Rev. **B 33**, 7375 (1986)
238. A.C. Ferreira, P.O. Holtz, B. Monemar, M. Sundaram, K. Campman, J.L. Merz, and A.C. Gossard, Phys. Rev. **B54**, 16994 (1996)
239. *Hydrogen in Crystalline Semiconductors*, ed. by S.J. Pearton, J.W. Corbett, and M. Stavola (Springer Series in Materials Science 16, 1992) p. 63, and references within
240. P.S. Nandhra, R.C. Newman, R. Murray, B. Pajot, J. Chevallier, R.B. Beall, and J.J. Harris, Semicond. Sci. Technol. **3**, 356 (1988)
241. S.J. Pearton, W.C. Dautremont-Smith, J. Lopata, C.W. Tu, J.C. Nabity, V. Swaminathan, M. Stavola, J. Chevallier, In:*GaAs and Related Compounds*, ed. by W.T. Lindley, Inst. Phys. Conf. Ser. 83, 289 (Inst. Phys., Bristol, UK 1987)
242. R. Murray, R.C. Newman, P.S. Nandhra, R.B. Beall, J.J. Harris, and P.J. Wright, MRS Proc. **104**, 340 (1988)
243. G.S. Jackson, N. Pan, M.S. Feng, G.E. Stillman, N. Holonyak, and R.D. Burnham, Appl. Phys. Lett. **51**, 1629 (1987)
244. B. Pajot, J. Chevallier, A. Jalil, and B. Rose, Semicond. Sci. Technol. 4, 91 (1989); N.M. Johnson, R.D. Burnham, R.A. Street, and R.C. Thornton, Phys. Rev. **B33**, 1102 (1986)
245. W.C. Dautrenart-Smith, MRS Proc. **104**, 313 (1988)
246. N. Pan, S.S. Bose, M.H. Kim, G.E. Stillman, F. Chambers, G. Devane, C.R. Ite, and M. Feng, Appl. Phys. Lett. **51**, 596 (1987)
247. J. Chevallier, B. Pajot, A. Jalil, R. Mostefaoui, R. Rahbi, and M.C. Boisy, MRS Proc. **104**, 281 (1988)
248. B. Monemar, P.O. Holtz, J.P. Bergman, Q.X. Zhao, C.I. Harris, A.C. Ferreira, M. Sundaram, J.L. Merz, and A.C. Gossard, Mat. Res. Soc. Symp. Vol. **325**, 19 (1994)
249. Y. Chung, D.W. Langer, R. Becker, and D. Look, IEEE Journal Electron Device, ED-32, 40 (1985)
250. P.E. Constant, Physica B **170**, 397 (1991)
251. L. Pavesi and P. Giannozzi, Phys. Rev. **B43**, 2446 (1991)
252. K.J. Chang and D.J. Chadi, Phys. Rev. Lett. **60**, 1422 (1988)
253. O. Nakajima, K. Nagata, and T. Makimura, Jpn. Journal of Appl. Phys. **31**, 1704 (1992)
254. J.P. Bergman, P.O. Holtz, B. Monemar, L. Lindström, M. Sundaram, J.L. Merz, and A.C. Gossard, In:*Proc. Intl. Conference on Defects in Semiconductors*, ICDS-18, Sendai, Japan, 199517 and C.I. Harris, M. Stutzmann, and K. Köhler, Mat. Sci. Forum, 117/118, 339 (1993)
255. I.A. Buyanova, A.C. Ferreira, P.O. Holtz, B. Monemar, K. Campman, J.L. Merz, and A.C. Gossard, Appl. Phys. Lett. **68**, 1365 (1996)
256. Q.X. Zhao, B.O. Fimland, U. Södervall, M. Willander, and E. Selvig,, Appl. Phys. Lett. **71**, 2139 (1997)
257. Q.X. Zhao, U. Södervall, M. Willander, B.O. Fimland, D. Crawford, E. Selvig, P.O. Holtz, M. Karlsteen, and E. Sveinbjornsson, MRS Proceeding Vol. **513**, 253 (1998)
258. A.P. Jacob, Q.X. Zhao, M. Willander, T. Baron, and N. Magnea, J. Appl. Phys. **90**, 2329 (2001)

Index

Springer Series in
MATERIALS SCIENCE

Editors: R. Hull   R. M. Osgood, Jr.   J. Parisi   H. Warlimont

Springer Series in
MATERIALS SCIENCE

Editors: R. Hull   R. M. Osgood, Jr.   J. Parisi   H. Warlimont

Printing: Strauss GmbH, Mörlenbach
Binding: Schäffer, Grünstadt

CPSIA information can be obtained at www.ICGtesting.com
Printed in the USA
LVOW100449150512

281757LV00005B/8/A